ELECTRONIC EQUIPMENT MAINTENANCE

AUTHOR

Dr. Chanchal Sharma

Professor,

Dept. of Electronics and Communication Engineering (ECE)

Arya College of Engineering and IT

Jaipur (Rajasthan)

REVIEWER

Prof. Pradeep Chindhi

Assistant Professor,

Dept. of Electrical Engineering (EE)

Sant Gajanan Maharaj College of Engineering

Mahagaon (Maharashtra)

KHANNA BOOK PUBLISHING CO. (P) Ltd.

4C/4344, Ansari Road, Darya Ganj, New Delhi, India

All India Council for Technical Education

Nelson Mandela Marg, Vasant Kunj, New Delhi, India

Dear Readers,

To prevent the piracy, this book is secured with HIGH SECURITY HOLOGRAM on the front title cover. In case you don't find the hologram on the front cover title, please write us to at contact@khannabooks.com or whatsapp us at +91-99109 09320 and avail special gift voucher for yourself.

Specimen of Hologram on front Cover title:

Moreover, there is a SPECIAL DISCOUNT COUPON for you with EVERY HOLOGRAM.
How to avail this SPECIAL DISCOUNT:
Step 1: Scratch the hologram
Step 2: Under the scratch area, your "coupon code" is available
Step 3: Logon to www.khannabooks.com
Step 4: Use your "coupon code" in the shopping cart and get your copy at a special discount
Step 5: Enjoy your reading!

ISBN: 978-93-5538-637-3
Book Code: DIP469EN

Electronic Equipment Maintenance
by Dr. Chanchal Sharma

First Edition: 2026

Originally Published by:
All India Council for Technical Education, Ministry of Education, GOI

Exclusively Published & Distributed by:
Khanna Book Publishing Co. (P) Ltd.
4C/4344, Ansari Road, Daryaganj,
New Delhi (INDIA)

Visit us at: **www.khannabooks.com**
Write us at: **contact@khannabooks.com**

To view complete list of books, Please scan the QR Code: KPH

Printed in Bharat.

प्रो. टी. जी. सीताराम
अध्यक्ष
Prof. T. G. Sitharam
Chairman

सत्यमेव जयते

आज़ादी का अमृत महोत्सव

अखिल भारतीय तकनीकी शिक्षा परिषद्
(भारत सरकार का एक सांविधिक निकाय)
(शिक्षा मंत्रालय, भारत सरकार)
नेल्सन मंडेला मार्ग, वसंत कुंज, नई दिल्ली–110070
दूरभाष : 011–26131498
ई–मेल : chairman@aicte-india.org

ALL INDIA COUNCIL FOR TECHNICAL EDUCATION
(A STATUTORY BODY OF THE GOVT. OF INDIA)
(Ministry of Education, Govt. of India)
Nelson Mandela Marg, Vasant Kunj, New Delhi-110070
Phone : 011-26131498
E-mail : chairman@aicte-india.org

FOREWORD

Engineers are the backbone of any modern society. They are the ones responsible for the marvels as well as the improved quality of life across the world. Engineers have driven humanity towards greater heights in a more evolved and unprecedented manner.

The All India Council for Technical Education (AICTE), have spared no efforts towards the strengthening of the technical education in the country. AICTE is always committed towards promoting quality Technical Education to make India a modern developed nation emphasizing on the overall welfare of mankind.

An array of initiatives has been taken by AICTE in last decade which have been accelerated now by the National Education Policy (NEP) 2020. The implementation of NEP under the visionary leadership of Hon'ble Prime Minister of India envisages the provision for education in regional languages to all, thereby ensuring that every graduate becomes competent enough and is in a position to contribute towards the national growth and development through innovation & entrepreneurship.

One of the spheres where AICTE had been relentlessly working since past couple of years is providing high quality original technical contents at Under Graduate & Diploma level prepared and translated by eminent educators in various Indian languages to its aspirants. For students pursuing 2nd year of their Engineering education, AICTE has identified 87 books, which shall be translated into 12 Indian languages - Hindi, Tamil, Gujarati, Odia, Bengali, Kannada, Urdu, Punjabi, Telugu, Marathi, Assamese & Malayalam. In addition to the English medium, books in different Indian Languages are going to support the students to understand the concepts in their respective mother tongue.

On behalf of AICTE, I express sincere gratitude to all distinguished authors, reviewers and translators from the renowned institutions of high repute for their admirable contribution in a record span of time.

AICTE is confident that these outcomes based original contents shall help aspirants to master the subject with comprehension and greater ease.

T.G. Sitharam

(Prof. T. G. Sitharam)

ACKNOWLEDGEMENT

The authors are grateful to the authorities of AICTE, particularly Prof. T. G. Sitharam, Chairman; Prof. Abhay Jere, Vice-Chairman; Prof. Rajive Kumar, Member-Secretary and Dr.Amit Kumar Srivastava, Director, Faculty Development Cell for their planning to publish the books on "Electronic Equipment Maintenance"We sincerely acknowledge the valuable contributions of the reviewer of the book Prof. Pradeep Chindhi (Assistant Professor) from Sant Gajanan Maharaj College of Engineering Mahagaon for making it students' friendly and giving a better shape in an artistic manner.

I convey thanks to Er.Anurag Agarwal (Chairman),Dr.Arun Arya (Principal), Prof.Rahul Srivastav from Arya College of Engineering and IT,Jaipur for their valuable suggestion to help me to bring this book in its final shape within the given time.

I would also like to give special thanks to my husband,my mother and whole family for their endless love and continuous support throughout my life.

This book is an outcome of various suggestions of AICTE members, experts and authors who shared their opinion and thought to further develop the engineering education in our country. Acknowledgement are due to the contributors and different workers in this field whose published books, review articles, papers, photographs, footnotes, references and other valuable information enriched us at the time of writing the book.

Dr. Chanchal Sharma

PREFACE

The book titled "Electronic Equipment Maintenance" is an outcome of the rich experience of our teaching of basic troubleshooting method of instruments The generation has altered the way electronic products are designed, fabricated, and maintained. For that reason, the personnel and the provider bench must preserve pace with these trends. This has created the want for a book that maintains in mind the generation trends and the growing importance of maintenance engineering. Historically maintenance engineering has no longer been a favored career of most of bright students. They discover layout and advertising and marketing jobs more challenging. However, with the liberalization of the economic system and globalization of activities inside the country, the sector of renovation engineering is increasing quite fast. it's been realized that at present, extra than ever before, the reliability of a machine could have a direct and on-the-spot impact on the profitability of an operation or the efficiency of a plant, health center, or service business enterprise. Hence, renovation engineering nowadays offers similarly right tough opportunities to service engineers as other desired expert fields. Technology is an ever-transferring goal. It's far nicely understood now that the times of troubleshooting a system simplest with an oscilloscope and multimeter and an indistinct idea about the hardware are over. These days, unless you have an in-depth service and protection manual and the proper kind of test gadget, you may be some distance from any success in troubleshooting and repairing of system. Most of the system problems can nevertheless be placed with a digital multimeter and oscilloscope if provided with sufficient effort and time. Therefore, the proper check system turns into a need if one has to work with modern-day technology.

The goal of maintenance engineering ought to be to ensure the pinnacle best performance from every piece of hardware. On their very own these targets no longer simply get achieved. they are completed via organizing and working towards an amazing maintenance management device. The miniaturization of electronic devices like cell phones, camcorders, virtual diaries, etc., has taken location primarily with the advent and improvements in surface mount technology. restore and transforming on SMD-based revealed circuit boards, have been covered in the book. it is was hoped that the book will allow the service engineers to do their activity extra efficiently and that scholars will be well served with the aid of the exercise-orientated method of the book and broaden interest in pursuing renovation engineering as an effective career. In addition, besides some essential information for the users under the heading "Know More" we have clarified some essential basic information in the appendix and annexure section.

We sincerely hope that the book will inspire the students to learn and discuss the ideas behind troubleshooting the electronic devices and will surely contribute to the development of foundation knowledge of the subject. We would be thankful to all beneficial comments and suggestions which will contribute to the improvement of the future editions of the book. It gives us immense pleasure to place this book in the hands of the teachers and students. It was indeed a big pleasure to work on different aspects covering in the book.

Dr. Chanchal Sharma

OUTCOME BASED EDUCATION

For the implementation of an outcome based education the first requirement is to develop an outcome based curriculum and incorporate an outcome based assessment in the education system. By going through outcome based assessments, evaluators will be able to evaluate whether the students have achieved the outlined standard, specific and measurable outcomes. With the proper incorporation of outcome based education there will be a definite commitment to achieve a minimum standard for all learners without giving up at any level. At the end of the programme running with the aid of outcome based education, a student will be able to arrive at the following outcomes:

Programme Outcomes (POs) are statements that describe what students are expected to know and be able to do upon graduating from the program. These relate to the skills, knowledge, analytical ability attitude and behaviour that students acquire through the program. The POs essentially indicate what the students can do from subject-wise knowledge acquired by them during the program. As such, POs define the professional profile of an engineering diploma graduate.

National Board of Accreditation (NBA) has defined the following seven POs for an Engineering diploma graduate:

PO1. Basic and Discipline specific knowledge: Apply knowledge of basic mathematics, science and engineering fundamentals and engineering specialization to solve the engineering problems.

PO2. Problem analysis: Identify and analyses well-defined engineering problems using codified standard methods.

PO3. Design/ development of solutions: Design solutions for well-defined technical problems and assist with the design of systems components or processes to meet specified needs.

PO4. Engineering Tools, Experimentation and Testing: Apply modern engineering tools and appropriate technique to conduct standard tests and measurements.

PO5. Engineering practices for society, sustainability and environment: Apply appropriate technology in context of society, sustainability, environment and ethical practices.

PO6. Project Management: Use engineering management principles individually, as a team member or a leader to manage projects and effectively communicate about well-defined engineering activities.

PO7. Life-long learning: Ability to analyse individual needs and engage in updating in the context of technological change

COURSE OUTCOMES (CO)

Students will be Able to:

CO-1: Understand the procedure of troubleshooting electronic equipment

CO-2: Test passive components such as Resistors, capacitors, and inductors

CO-3: Identify causes of failures in semiconductor devices and fault diagnosis in OPAMP circuits

CO-4: Apply knowledge of digital IC testing in fault diagnosis of digital circuits

CO-5: Repair surface mount assemblies and PCBs

CO-6: Evaluate electronic equipment's working condition

Mapping of Course Outcomes with Programme Outcomes to be done according to the matrix given below:

Course Outcomes	EXPECTED MAPPING WITH PROGRAMME OUTCOMES (1- Weak Correlation; 2- Medium correlation; 3- Strong Correlation)						
	PO-1	PO-2	PO-3	PO-4	PO-5	PO-6	PO-7
CO-1	3	3	1	3	2	2	3
CO-2	2	3	2	3	2	1	2
CO-3	1	2	3	3	2	1	2
CO-4	2	3	2	3	2	2	3
CO-5	2	3	2	3	3	2	2

GUIDELINES FOR TEACHERS

To implement Outcome Based Education (OBE) knowledge level and skill set of the students should be enhanced. Teachers should take a major responsibility for the proper implementation of OBE. Some of the responsibilities (not limited to) for the teachers in OBE system may be as follows:

- Within reasonable constraint, they should manoeuvre time to the best advantage of all students.
- They should assess the students only upon certain defined criterion without considering any other potential ineligibility to discriminate them.
- They should try to grow the learning abilities of the students to a certain level before they leave the institute.
- They should try to ensure that all the students are equipped with the quality knowledge as well as competence after they finish their education.
- They should always encourage the students to develop their ultimate performance capabilities.
- They should facilitate and encourage group work and team work to consolidate newer approach.
- They should follow Blooms taxonomy in every part of the assessment.

Bloom's Taxonomy

Level	Teacher should Check	Student should be able to	Possible Mode of Assessment
Creating	Students ability to create	Design or Create	Mini project
Evaluating	Students ability to Justify	Argue or Defend	Assignment
Analysing	Students ability to distinguish	Differentiate or Distinguish	Project/Lab Methodology
Applying	Students ability to use information	Operate or Demonstrate	Technical Presentation/ Demonstration
Understanding	Students ability to explain the ideas	Explain or Classify	Presentation/Seminar
Remembering	Students ability to recall (or remember)	Define or Recall	Quiz

GUIDELINES FOR STUDENTS

Students should take equal responsibility for implementing the OBE. Some of the responsibilities (not limited to) for the students in OBE system are as follows:

- Students should be well aware of each UO before the start of a unit in each and every course.
- Students should be well aware of each CO before the start of the course.
- Students should be well aware of each PO before the start of the programme.
- Students should think critically and reasonably with proper reflection and action.
- Learning of the students should be connected and integrated with practical and real life consequences.
- Students should be well aware of their competency at every level of OBE.
- Students should be well aware of their competency at every level of OBE.

ABBREVIATIONS AND SYMBOLS

List of Abbreviations

General Terms			
Abbreviations	**Full form**	**Abbreviations**	**Full form**
TTL	Transistor Transistor logic	CMRR	Common Mode Rejection Ratio
MOV	Metal Oxide Varistors	SR	Slew Rate
LDR	Light Dependent Resistors	ECL	Emitter Coupled Logic
SMD	Surface-Mount resistors	IC	Integrated Circuit
POT	Potentiometer	SIP	Single Inline Package
DMM	Digital Multimeter	CFP	Ceramic Flat Pack
LED	Light Emitting Diodes	QCFP	Quad Flat Pack
TD	Tunnel Diode	SOP	Small -Outline Package
BJT	Bipolar Junction Transistor	CSOP	Ceramic Small Outline Package
FET	Field Effect Transistor	DSOP	Dual Small Outline Package
JFET	Junction Field Effect Transistor	HSOP	Thermally Enhanced Small
MOSFET	Metal Oxide Semiconductor	SSOP	Outline Package
	Field Effect Transistor	TSOP	Shrink Small Outline Package
OM-AMP	Operational Amplifier	TVSOP	Thin-Small Outline Package
HTSSO	Thermally enhanced thin shrink		Thin Very Small Outline Package
	small-outline package	MSOP	Micro Small Outline Package
QSOP	Quarter Small Outline Package	HSSOP	Thermally-Enhanced Shrink
SOICW	Small Outline Integrated Circuit		Small Outline Package
	Wide	SOIC	Small Outline Integrated Circuit
PSON	Plastic Small Outline No Lead	Mini	Mini-Small Outline Integrated
	Package	STOIC	Circuit
		PSOP	Plastic Small Outline Package

List of Units and Symbols

Symbols	Description	Symbols	Description
______	It is the symbol that is used to represent a wire.	I	Electric Current
AC	Alternating Current	I_C	Collector Current
Hz	Hertz	μV	Microvolt
VDD	Drain Supply Voltage	pF	Picofarad
β (beta)	Current Gain	Ipp	Peak-to-Peak Current
BW	Bandwidth	kHz	Kilohertz
C	Capacitor	μA	Milliampere
DC	Direct Current	μA	Microampere
f	Frequency	μH	Microhenry
fr	Resonant Frequency	Ω	Ohms
H	Henry	τ	Time Constant

LIST OF FIGURES

LIST OF TABLES

CONTENTS

1 Fundamental Troubleshooting Procedures Inside an Electronic Equipment

UNIT SPECIFICS

Through this unit we have discussed the following aspects:

- Reading and drawings electronic block diagram
- Dis-assembly and re-assembly of electronic equipment.
- Inappropriate operating conditions,
- Grounding systems in electronic equipment,
- Temperature sensitive intermittent problems;
- Problems related to nature of faults, fault location and operating procedure,

The discussion of these issues' real-world applications is meant to inspire more inquiry, imagination, and problem-solving skills,assignments and, a list of references, and suggested readings are given in the unit so that one can go through them for practise, in addition to a large number of multiple choice questions and questions of short and long answer types marked in two categories following lower and higher order of Bloom's taxonomy. It's worth noting that QR codes, which may be scanned for further information, have been included in various parts so that readers can learn more about the themes that interest them.

Related exercises are followed by a ''Know More" section that expands on the material covered in the exercises. This appendix has been thoughtfully constructed to ensure that the additional material included here is useful to readers of this book. This part focuses on the topic's origins, provides instances of relevant facts and analogies, traces its historical development while emphasizing key observations and findings, provides timelines that span the evolution of the relevant subjects from their earliest stages all the way up to the present day, and discusses the topic in terms of its relevance to our everyday lives and the ways in which it may be used in the workplace.

RATIONALE

This fundamental unit on Troubleshooting procedures helps students to get a basic idea about the maintenance of electronic equipment and the nature of faults. It explains the concept of reading and drawing block diagrams and wiring diagrams of electronic equipment and its completeness in describing the right way of operating any equipment.

All these fundamental aspects are relevant to start the troubleshooting of equipment properly. It then explains the Fault finding aids and their completeness in describing the service and maintenance manuals and instruction manuals; All these are discussed at length to develop the subject. Some related problems related with an extension to testing of Measuring instruments, which can help further get a clear idea of the concern topics on Troubleshooting Procedures. Over the past few years, the electronic sector has seen a revolutionary transformation. The development, assembly, and servicing of electronic goods have all been affected by technological advancements. Since maintenance engineering is becoming increasingly vital, it is imperative that both staff and the service bench stay up with the times. The purpose of maintenance engineering is to keep all equipment in peak working condition. To accomplish these tasks, a faultless maintenance management system must be developed and implemented.

PRE-REQUISITES

Physics: Electrostatics, Magnetic Effects of current and magnetism, Electromagnetic Waves, Semiconductor Electronics (Class XII)

Automotive: Measuring & service Equipment Steering system Suspension system Transmission and Final Drive system (Class XII)

UNIT OUTCOMES

List of outcomes of this unit is as follows:

U1-O1: Discuss the concept of reading and drawing electronic block and wiring diagrams.

U1-O2: Describe the troubleshooting of fault and appropriate operating procedure

U1-O3: Explain dis-assembly and assembly of measuring instruments

U1-O4: Realize the role of semiconductor devices in equipment manufacturing

U1-O5: Apply the procedure of Grounding in Electronic Equipment

Unit-1 Outcome	EXPECTED MAPPING WITH COURSE OUTCOMES (1 - Weak Correlation; 2 - Medium Correlation; 3 - Strong Correlation)					
	CO-1	CO-2	CO-3	CO-4	CO-5	CO-6
U1-O1	2	3	3	-	3	1
U1-O2	1	1	2	3	1	2
U1-O3	3	2	3	2	3	1
U1-O4	2	3	-	1	1	3
U1-O5	3	3	3	-	3	2

1.1 INTRODUCTION

Any electronic equipment or a system comprises sub-systems like different blocks to evaluate specific functions. These sub-systems are the summation of electronic circuits. These circuits may be electronic/electrical or electromechanical parts. The system is a failure if any equipment component is not working efficiently. Failure means "the inability of a system to perform expected outcomes." If the continuous performance of equipment is needed, then regular maintenance is required. For effective troubleshooting of faulty equipment, technicians should have the correct information on the instrument's internal components, circuit diagram, and front panel control. The equipment's maximum performance depends upon proper installation, routine checkups repairing and replacement parts, and suitable servicing activities. This chapter emphasizes the basic requirement for troubleshooting techniques and safety methods by which isolation of faults possible

1.2 READING AND DRAWING SCHEMATIC DIAGRAM

A drawing of any electrical or electronic circuit is a graphical representation of connections and components. It can be a wiring diagram, electrical diagram, elementary diagram, block representation, etc., for designing, building, and troubleshooting purposes, and correctly reading and understanding of schematics diagrams is an essential skill for any profession.

1.2.1 Block Diagram

When in all types of equipment, all interacting components or elements respond to applied inputs for producing appropriate output, that equipment can be considered a System. When analyzing a system, it is sometimes essential to break it down into smaller, more manageable pieces, or subsystems. Each subsystem would be represented by a functional block, and the whole would be shown by a functional block diagram of the equipment. It is having the necessary inputs and outputs to carry out a specified operation as shown in Figure 1.1.

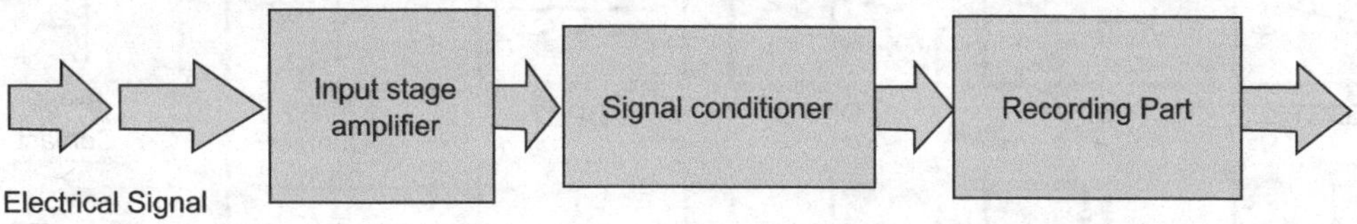

Fig. 1.1: Representation of block diagram.

It depicts the various components of the electronic device. An input stage amplifier is used in a basic recorder to record electrical signals from any source. After the amplifier phase, a signal conditioner is used to prepare the output for use by the rest of the system. Depending on the nature of the assignment, the signal conditioner may take the form of an amplifier, a signal modifier, or a signal attenuator. The signal flow channel is shown as a series of interconnected blocks, each of which represents a different system variable. When troubleshooting a defect, it is necessary to focus on a single conditioner block rather than the entire circuit. The block diagram method is useful for elucidating the interconnections between components and facilitating repairs by targeting the specific dysfunctional building block that needs fixing.

1.2.2 Circuit Diagram

The graphical representation of internally connected components of any equipment is known as a circuit diagram. The circuit board and the schematic will often include an assembly number for each completed piece of hardware. Symbols representing each part of the circuit are laid out in the correct order in the circuit diagram, which serves to explain how the circuit operates. With so many different rules governing the symbols used for components, it's important to know which one has been used before attempting to decipher a circuit design. The combination of these symbol is sometimes referred to as a schematic drawing.

Following are the guidelines for drawing diagrams the main features of these guidelines are:

- Across the page, the signal travels from left to right, with inputs on the left and outputs on the right.
- The voltage should rise from the bottom to the top of the page. In Fig. The +5V supply is depicted vertically in Figure 1.3, whereas the ground pin is depicted horizontally.
- For each package, use the 'unit number' convention to designate an Ul, with the internal gates indicated by a letter suffix (U1A, U1B, etc.). All power connections can be shown on a single conventional gate (pin 7 and pin 14).
- Power connections are often specified, however they should be added ahead of time for a more comprehensive schematic.
- It is recommended that any work on troubleshooting begin with a comprehensive and easily understandable engineering schematic diagram that depicts all relevant components and their connections.

Know more about circuit and wiring diagram

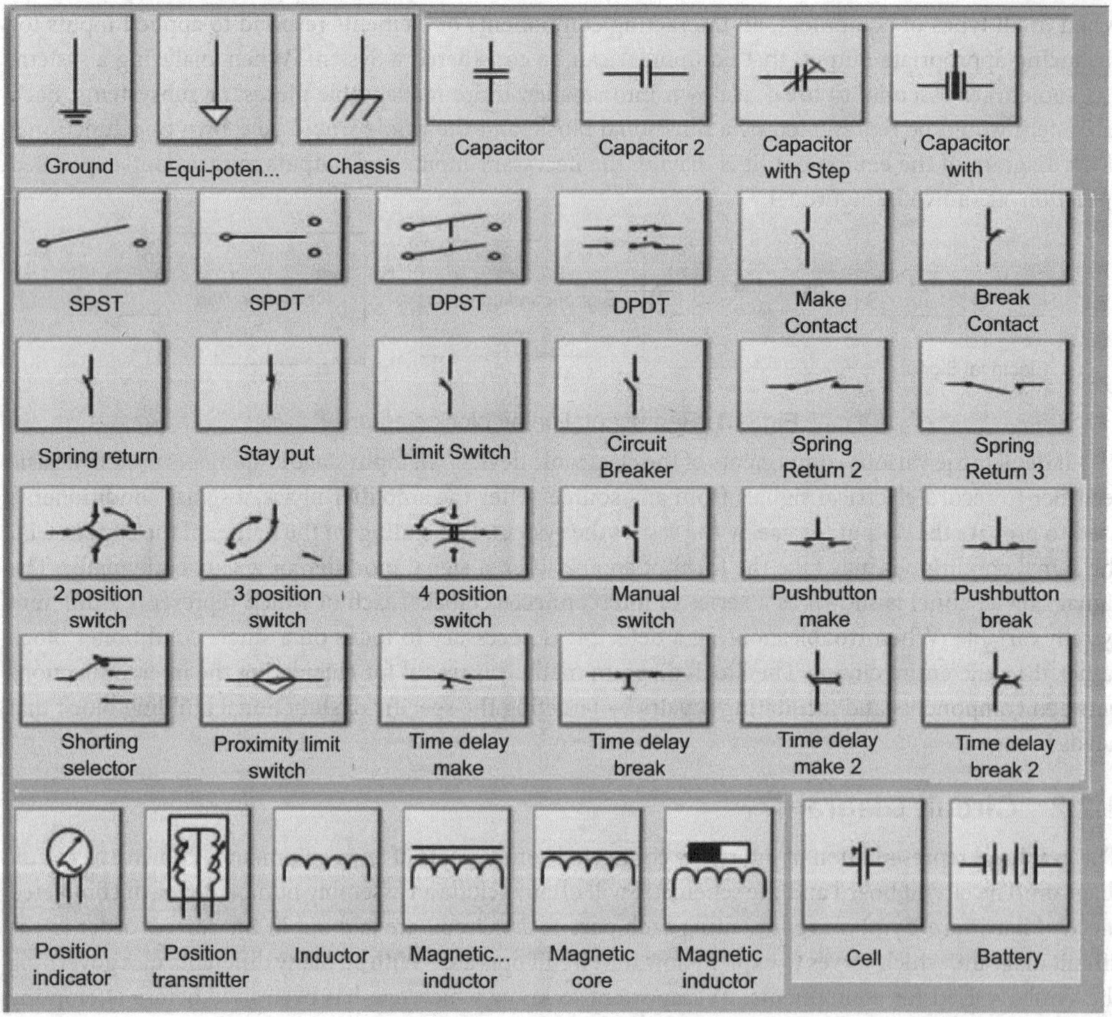

Fig. 1.2: Special symbols used on circuit diagrams

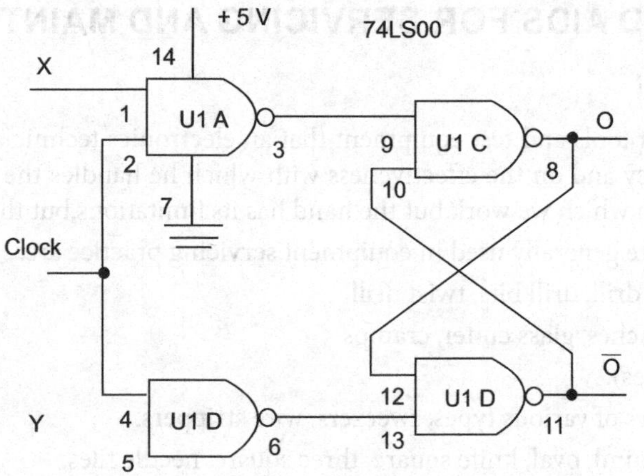

Fig. 1.3 : Example of an engineering schematic

1.2.3 Wiring Diagram

The purpose of a wiring diagram is to demonstrate how the wires are connected on a circuit board or in the building. And these connections have no further information except the signal names. So it concludes that the process will be complex or impossible if we use only the wiring technique for troubleshooting.

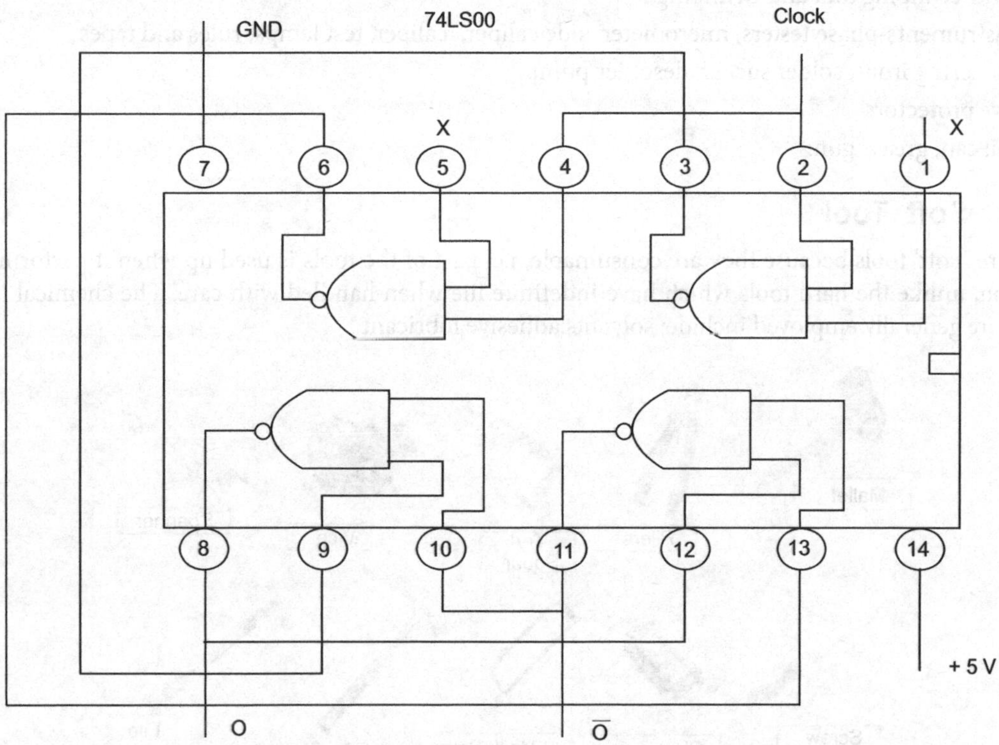

Fig. 1.4 : Representation of a wiring diagram

1.3 TOOLS AND AIDS FOR SERVICING AND MAINTENANCE

1.3.1 Hand Tool

The type and variety of tools and test equipment that an electronics technician possesses have a direct bearing on his efficiency and on the effectiveness with which he handles the job.The human hand is a unique instrument with which we work,but the hand has its limitations.but the hand has its limitations.

- The tools which are generally used in equipment servicing practice are:
- Power drill ,Hand drill, drill bits, twist drill
- Taps and dies, punches, glass cutter, cramps
- Spanners (wrenches).
- Pliers-cir-clip pliers of various types, tweezers, wire strippers.
- File-round, half round, oval, knife square, three square, needle files.
- Ball pen ,Hammer-clown,cross peen
- Screw drivers of various sizes
- Bench grinder, oil stone.
- Bench vice
- Brushes-soft and wire type
- Saws-hacksaw, piercing saw
- Anti-crimping tool and Crimping
- Instruments-phase testers, micrometer ,side caliper, caliper, test lamps, rules and tapes
- Soldering irons, solder sucker, desolder pump
- Eye protectors
- Oil can, grease gun.

1.3.2 Soft Tool

They are `soft' tools because they are consumable, i.e. part of the tools is used up when it performs its function, unlike the hard tools which have indefinite life when handled with care. The chemical tools which are generally employed include: solvents,adhesive,lubricant

Fig. 1.5: Different types of hand tools

1.4 DISASSEMBLY AND REASSEMBLY OF EQUIPMENT

Disassembly and reassembly of equipment is a common troubleshooting technique. However, disassembly and reassembly need a careful attention, especially if getting to the internal workings of the apparatus is difficult. Use these instructions to take apart and put back together your apparatus with ease.

1.4.1 Disassembly

Disassembling anything is best done with the help of the manufacturer-supplied service manual or other documentation. Unfortunately, there are situations when this data isn't included in the product manual. When it comes to unlocking their products, certain brands appear to take great satisfaction in being very cryptic. Opening the equipment without causing any damage is always the hardest aspect of repairing it. Many factors contribute to the safety of electronic devices, including as:

Screws

The Phillips head is the standard for screws, even in older, less advanced tools. An evenly engraved arrow on the casing usually denotes access panels for removing screws. The trim panel pops out to the apparent screw holes, the trim. Some may only be accessible when we open a battery or cassette compartment.

Hidden Screws

These will need a fancy label to be peeled off or a plug to be pried out. The rubber may break out sometimes, exposing the screw holes below. If there is a sticker on the label, you can find a hidden screw by rubbing your finger over it.

Snaps

To reveal the locations of snaps, you can press gently or forcibly, with the help of a screwdriver, sometimes pops out the surface in one place with a knife or screwdriver to unlock the covers.

Fused Casings

To protect the internal components of an AC wall adapter, a fused plastic case is often employed. On other devices, the epoxy is potted all the way to the top. To make repairs, you may disassemble some of them using a hacksaw blade and then use plastic electrical tape to put them back together.

Plastic Catches

The use of Plastic catches is to secure LCD housings. Sometimes screws are used in inventive ways, such hiding them in between the stitches. Be patient as you try to locate the catches and screws. Don't try to open forcefully any component of machinery unless you have to and can't think of any other way to get inside. The covers are simple to remove once you know how they are fastened. If they become stuck, check for any remaining screws or snap.

1.4.2 Reassembly

After equipment repair, all parts should be reassembled correctly together without any external force. You can easily reassemble if you remember how to disassemble the pieces. So the knowledge of circuit diagrams with wiring and components is necessary for reassembling any equipment.

1.5 EQUIPMENT FAILURES AND CAUSES OF EQUIPMENT FAILURES

Any piece of machinery is considered to have failed when it is no longer sufficient to carry out its intended function, when its characteristics have changed, and when it is no longer able to complete the process to the needed level. Partial failures occur when some characteristics or parameters go beyond the allowed range. Still, if the qualities vary beyond the stipulated limits, causing a total degradation of the required procedure, we speak of a full failure. Failures may sometimes be anticipated by examination (gradual loss), but they can sometimes occur suddenly, which is difficult to foresee through normal inspections. A catastrophic failure occurs when there is an abrupt or total change in the operation of the device. The cause might be a temporary open circuit or a permanent short circuit. An open circuit in a wire-wound potentiometer or a short circuit in a capacitor are two common examples of this kind of failure. The only way to fix this kind of failure is to either remove the offending part or replace it. Failure may also happen progressively and in stages. Degradation failure refers to a failure that causes some functionality to be lost without completely crippling the system. These kinds of failures are more common in analogue signals, especially when there is noise or disturbance present, and are far less common in digital ones. When integrated circuits were first used for complicated operations, they greatly reduced the risk of deterioration. They are less common in non-critical applications, but nonetheless common in machinery that must function in harsh environments. Different factors might lead to equipment failure. They may be broken down into the following categories:

1.5.1 Poor Design

Some of equipment having poor design due to

(a) Materials which is using that is not appropriate

(b) Lack of basic Knowledge of the component and schematic diagrams

(c) Poor layout and parts of mechanical panels

(d) Excessive heat inside the equipment

(e) Insufficient specifications like reliability and performance

1.5.2 Production Deficiencies

1. Goods, sampling tests, and other components lack in inspection
2. Due to prolonged storage period and insufficient method of storage of components

3. Sometimes technician lack in the necessary motivation, expertise, and commitment to do their jobs effectively.
4. Technician not having correct and most influential production techniques due to the absence of training programs
5. Using low-quality machinery and tools in production
6. Poor working conditions, include badly illuminated and dusty assembly rooms that lack sufficient ventilation and lighting.
7. Inadequate inspection and testing of final items
8. Failure to properly conduct environmental tests, such as cycling temperatures and running machinery at abnormally high temperatures for prescribed amounts of time.

1.5.3 Careless Storage and Transport

1. Unused machinery lying in warehouses from long time before being sent out to a client.
2. Corrosion and other damage could occur to the equipment if it wasn't properly packaged.
3. When equipment transfer from the factory to the end user, it is subjected to a lot of mechanical shocks and vibrations.

1.5.4 Inappropriate Conditions during Working Life

1. A hostile work environment, including a lack of air conditioning and dust-free spaces;
2. Accidents caused by misuse, abuse, or disregard for equipment safety instructions, cautions, and warnings
3. Inaccurate results and inefficient operation of that equipment.
4. Inadequate after-sales support from manufacturers.
5. Fluctuations in mains voltage (a issue that frequently occurs in developing countries)
6. Using a equipment after its intended lifespan has ended;
7. Due to a lack of routine and decided maintenance.

1.6 FAULT MANAGEMENT

1.6.1 Nature of Faults

1. The nature of faults may vary from simple mechanical faults to complex faults in electronic circuitry. Some results shows that about 30% of the faults are minor in nature, such as loose sockets ,a blown up fuse, broken , sticky panel meters, or defective connectors, or shorted power cord not proper ground leads ,etc.Inappropriate environments effects equipment to get heated up and gives instability and drift, Most of the presence of disturbing elements in the surroundings of the equipment cannot be considered as faults in the system, but sometimes they could be the cause of big anxiety.

2. About 20% of the faults are of a very common nature. These are mostly mechanical faults. Such faults as loose mechanical fixtures, and stylus of recorders, mounting of printed circuit boards in connectors and loose valve, knobs, handles, power sockets transistor and IC base contacts. Sometimes, These faults may be due to drained out dry batteries, uncharged Ni-Cd cells, burnt-out pilot lamps, defective sensors etc about 30% to 40% of the equipment develops faults which are neither simple nor common in nature
3. These faults are specific to the equipment concerned and include burnt transformers, erratic performance due to instability and failure of one or more stages in the equipment due to voltage fluctuations or carelessness of the user.Another 10% of the faults which are observed in the equipment are chronic in nature which become a headache for both the service engineer and the user. These faults can be termed as `repetitive faults' and occur due to either poor designing, or the use of substandard components in its manufacture, or poor quality control or the prolonged use of equipment beyond its expected working life.Another problems in electronic equipment is power supply consisting of damage due to over- heating and tripped over current protection devices

Active components tend to fail with greater regularity than passive devices, due to their greater complexity. Semiconductor devices are notoriously prone to failure due to electrical transient (voltage/ current surge) overloading and thermal (heat) overloading.Passive components are the most rugged of all. However, the following failure probabilities have been generally noticed .

It is usually simple to get to the issue place if you use a methodical and logical approach while attempting to locate a mistake. However, the technician's understanding of the apparatus and aptitude for troubleshooting are crucial in determining where a specific defect lies. In order to effectively troubleshoot a system, one must be intimately conversant with the components it uses. Take the hypothetical scenario of not being able to investigate the workings of a microcomputer, decipher its indicators, understand its printouts, or evaluate its software. In such instance, isolating a problem inside that piece of machinery will be hard.

When troubleshooting, it's usually best to start with a quick analysis of the symptoms, which may narrow down the list of potential reasons. This is done in descending order of probability, and different short checks are often performed to rule out or confirm mutual deductions. The recommended overall approach for troubleshooting the equipment are shown in Figure 1.6. In order to effectively troubleshoot a system, one must be intimately conversant with the components Troubleshooting always begins with a preliminary investigation of the problem symptoms, from which several potential sources of failure may be narrowed down. These are ranked according to their probability, and many fast tests are often performed to rule out or confirm mutual reductions.

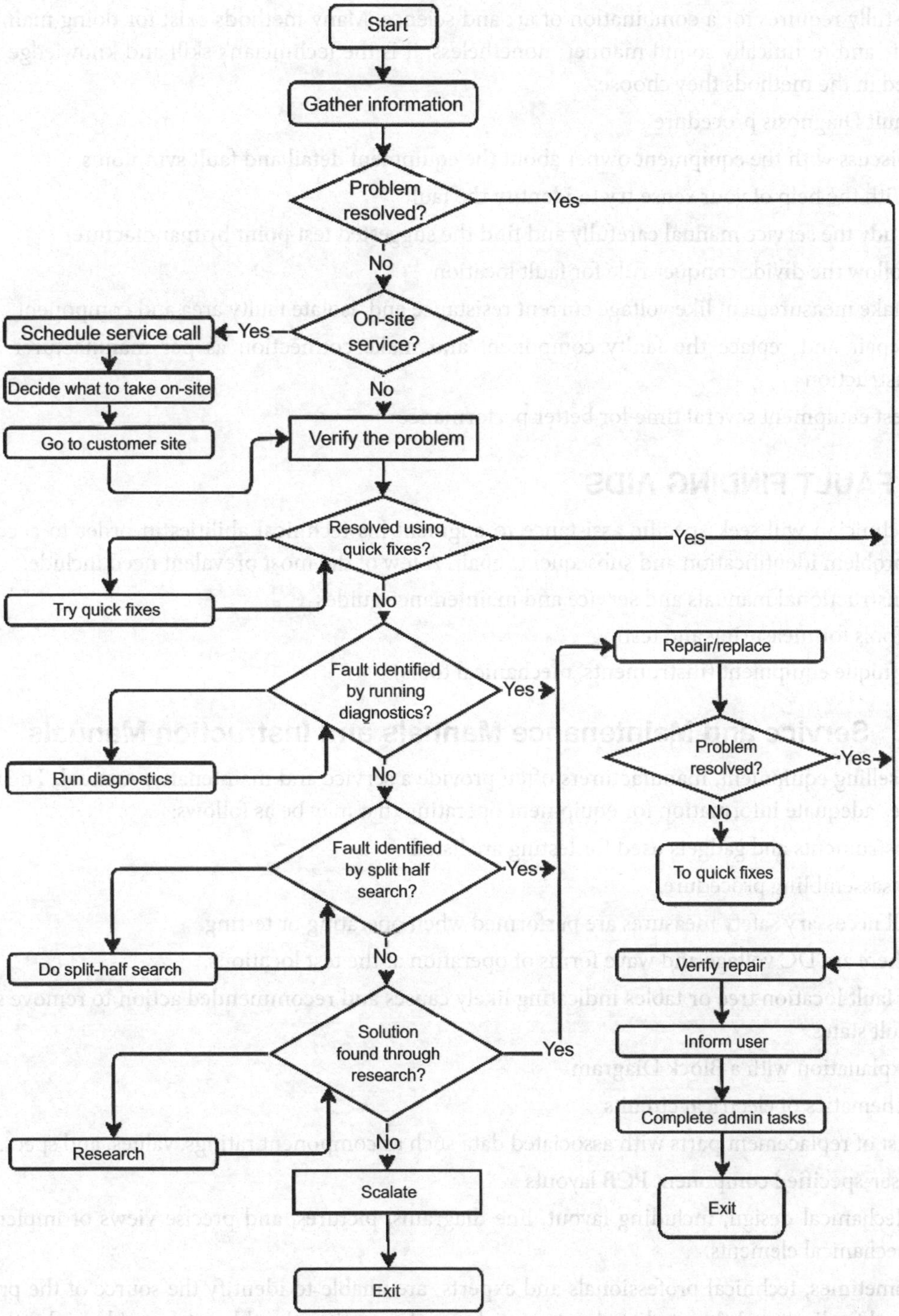

Fig. 1.6: Troubleshooting procedure

The recommended high-level approach to fixing problems is outlined in Figure 1.6. Solving a problem successfully requires for a combination of art and science. Many methods exist for doing maintenance in a safe and technically sound manner; nonetheless, it is the technician's skill and knowledge that are revealed in the methods they choose.

Fault Diagnosis procedure

- Discuss with the equipment owner about the equipment detail and fault symptoms
- With the help of your sense try to identify the fault
- Study the service manual carefully and find the suggested test point by manufacturer
- Follow the divide conquer rule for fault location
- Make measurement like voltage current resistance and isolate faulty area and component
- Repair and replace the faulty component and make connection as per manufacturer manual instruction
- Test equipment several time for better performance

1.7 FAULT FINDING AIDS

The technician will seek specific assistance to augment his technical abilities in order to accomplish rapid problem identification and subsequent repair. A few of the most prevalent need include:

(a) Instructional manuals and service and maintenance guides
(b) Tools for measuring and testing
(c) Unique equipment (instruments, mechanical tools)

1.7.1 Service and Maintenance Manuals and Instruction Manuals

When selling equipment, manufacturers often provide a service and maintenance manual. The manual provides adequate information for equipment operating that may be as follows:

- Instruments and gadgets used for testing are listed.
- Disassembling procedure.
- All necessary safety measures are performed when operating or testing.
- There are DC voltage and wave forms of operation at the test locations.
- A fault location tree or tables indicating likely causes and recommended action to remove specific fault states.
- Explanation with a Block Diagram
- Schematics of electrical circuits
- List of replacement parts with associated data such as component ratings, values, and specs.
- User-specified component PCB layouts
- Mechanical design, including layout, line diagrams, pictures, and precise views of implemented mechanical elements.

Sometimes, technical professionals and experts are unable to identify the source of the problem. Measured findings and the predicted outcomes from observation should match and be right, therefore it's time to check the service manual and attempt to understand both.

The availability of repair manuals is not always assured. Without a doubt, the technician will be able to make an effort at repairing the defect based on his or her previous expertise with other systems like it. However, if the system is unknown territory, pushing on with testing without a thorough understanding of the system's behaviour may often lead to erroneous findings and, in rare circumstances, the introduction of new defects. Therefore, before attempting to find a critical defect in the system, it is important to concentrate on the repair manual as much as possible. When a schematic design is unavailable, reverse engineering must be used to create the circuit diagrams. whether it known as a "instruction manual," "user manual," or "operator's manual," this document contains all the information you'll need to run the machine properly, from start-up and shut-down procedures to basic design principles, specifications, and installation methods. Your success in locating and resolving the defect will depend on how thoroughly you read the document. Operator maintenance instruction manuals are useful for performing preliminary diagnostics and creating a preventive maintenance programme; they are the backbone of preventive maintenance.

Know more about fault finding aid

1.8 TEST AND MEASURING INSTRUMENTS

In order to examine their findings, scientists in electronics and electrical measurement laboratories, research labs, manufacturing businesses, and disciplines dealing with materials and particles need a broad variety of testing and measuring equipment.The three primary diagnostic tools allow for identifying and fixing of faults. Testing and measuring equipment is extensively used in many different disciplines, including electronics and electrical measurement laboratories, research labs, industrial businesses, and material and particle analysis fields. These are the primary tools for detecting and fixing problems.

(a) Multi-meter:

The three most important electrical parameters may be measured using a multi metre: voltage, current, and resistance. In addition to its application in home wiring, electric motors, battery testing, and checking the output of power sources, it may also operate as an ohmmeter, voltmeter, and ammeter.

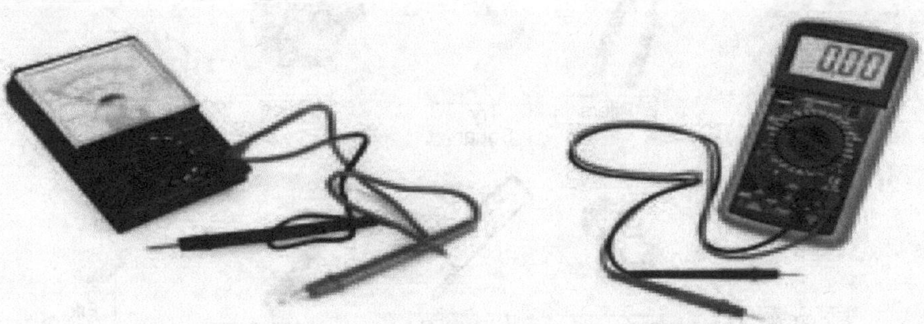

Fig.1.7: Analog and Digital Multi-meter

(b) Oscilloscope

An oscilloscope, also known as an oscillograph, a digital storage oscilloscope, or a cathode ray oscilloscope, is a part of electronic testing equipment that displays a two-dimensional plot of signals as a function of time for a given voltage.

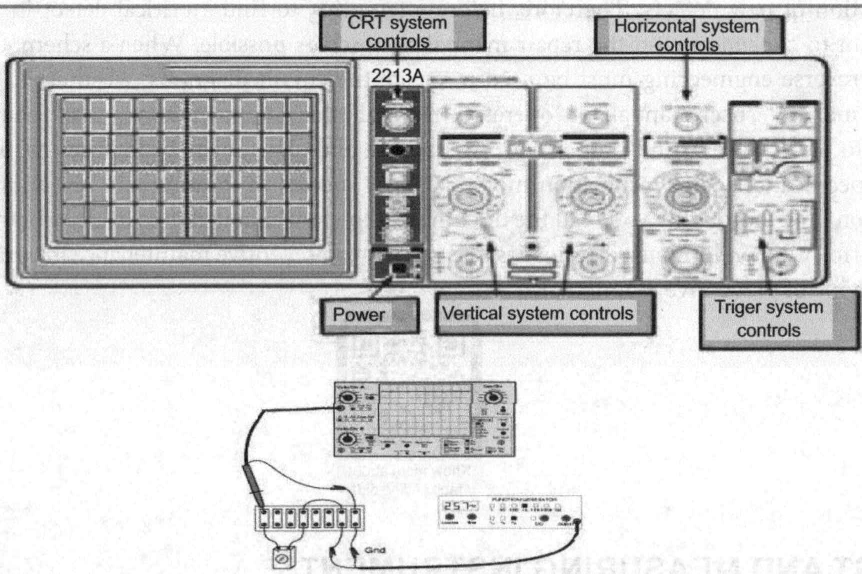

Fig:1.8: Measurement on oscilloscope using signal generator

(c) Signal generator, pulse generator, or function generator:

In order to effectively test, design, and repair electronic equipment, signal generators are important tools. The term "signal generator" refers to any of many kinds of electronic equipment that may be used to create arbitrary signals of a desired frequency, pitch, or other function.

(d) Special tools:

For routine servicing, some special hand tools and chemicals are required for troubleshooting the fault.

Fig.1.9: Types of Hand tools

With the help of measuring instruments, you can measure voltage amplitude-frequency and phase measurement, and on complex circuit functions, you must perform detailed waveform analysis to troubleshoot the fault.The test equipment generates signals and reads and checks the device's performance under examination.The variations in this response indicate the faults, and we can rectify them with the help of a suitable procedure. Equipment specific to testing parameters such as voltage, current, data transfer rates, power, etc.

1.9 TROUBLESHOOTING TECHNIQUES

1.9.1 Preliminary Observations

Preliminary inspections must be performed before any repairs are made to damaged machinery. In addition to ensuring your safety, these measures also make it easier to go closer to the source of the problem in record time.

Following are the safety precautions that should be taken:

(a) Make sure you check the panels on both sides of the equipment for important information from the manufacturer. Warning labels on panels use the following terminology:
   - The marking to property, including the equipment itself, shall be considered a danger if we obtain information about a personal harm hazard that is not immediately curable.
   - As one reads the marking, the word ''danger'' denotes a potential for personal damage. If high voltages exist on the equipment, then Danger signs will be marked at places

(b) Determine whether or whether the item needs a special power source. Some devices may run on batteries while others need to be plugged into the power. Specify whether the mains voltage is 110 V at 60 Hz or 220 V at 50 Hz, and the maximum voltage that may be drawn from the outlet. Identify the power lead (if the plug is missing) and the ground. The grounding conductor of an electrical cable provides the crucial grounding required for safe operation.

(c) Electronic devices are typically grounded by connecting the power cord's grounding conductor to a grounded metal object. To prevent electrical shock, make sure the power cable is plugged in firmly and correctly, and check all input and output terminal connections.

(d) Before powering up the apparatus for a preliminary inspection, read the service handbook carefully and be on the lookout for the following phrases:
   - Caution signs explain the kind of usage that might be harmful to machinery and other assets.
   - Statements of caution outline potential dangers that might cause serious harm to people or even death.
   - It also provides any further warnings or other relevant information.

(a) The service manuals typically include information on how to repair the equipment and how to gain access to the various circuit boards and components. These instructions must be followed, as opening the incorrect screws can dislodge the internal assemblies, causing the fragile parts to be effected.

(b) In order to prevent electric shock, turn off the power to the equipment before lifting the panels from the cabinet.

(c) Numerous places in certain equipment are energized at potentially lethal levels of voltage. When testing or servicing this equipment, it is imperative that no contact be made with any uncovered connections or components, even if the covers are removed. Some transistors have voltages on their casings, thus power must be cut off before cleaning or changing components.

(d) If you know very little about the contents of a potted assembly or component, don't risk further damaging the equipment by trying to crack it open.

(e) Do not make any kind of adjustment or conduct any kind of internal servicing while there are live voltages present, and first aid supplies should always be close at hand.

(f) Avoid using excessive force while handling equipment with painted metal surfaces since scratches are easily caused by this.

(g) Raw metal edges are visible when the equipment cover is about to open from the chassis, which might damage the work area. When working, make sure the chassis's corners and edges don't scuff the table or floor

(h) Some electronic devices are susceptible to a major collapse due to an uncontrolled static electricity discharge, especially those that utilize the CMOS logic family of integrated circuits. If there is a significant voltage differential between your body and the equipment, ground yourself by firmly grasping the chassis before touching any circuit boards.

(i) Handle all circuit boards by the edges. CMOS circuits run on nano ampere currents, and leakage routes from things like skin oils, grime, dust, etc. may lead to faulty circuit performance in certain devices. Workplace injuries may be avoided by always using PPE, hence it's imperative that workers always wear it.

(j) After taking off the protective cover, double check the screw tightness. We need to take a good, hard look at the positioning and connections on all of the printed circuit boards. All exterior wires should be inspected for damage, such as fractures and twisted sections. If there is significant damage, the cable should be replaced right away.Ultimately, the following three considerations should help you design a rational approach for troubleshooting.

1. Know your equipment: Without sufficient knowledge of the machinery to examine its functioning, understand its signs, and comprehend its printouts, issues will be difficult to resolve.
2. Think before you act: Avoid making decisions without prior approach to replace or remove components without first investigating probable reasons. A wrong approach to problem solving causes extra issues and can take an excessive amount of time.
3. Establish a general troubleshooting procedure: In order to update a current troubleshooting technique, one requires familiarity with the problem in operation as well as access to relevant information, instruments, replacement components, and time. The faulty boards are either fixed after some time or sent back to the manufacturer for servicing, and this process may involve fixing individual components or the entire board.

1.9.2 Troubleshooting Method

The troubleshooting methods described here are typical for fixing most electrical devices. The method(s) you use for troubleshooting will depend on the system you're working with and your level of expertise.

Functional Area Approach: Generators, amplifiers, generators, transmitters, converters, logic manipulators, and data storage devices are just some of the numerous components that may be included into an electronic system. The system fails to provide the desired results if the issue might occur in any of its possible functional domains. Therefore, it is crucial to do troubleshooting in order to locate the source of the problem and either fix it or replace the broken component. Finding a flaw with reasoning yields accurate results. It is possible to disentangle the source of a fault when it has been pinned down to a single operational subsystem. If you're having issues in a certain area of functionality, drawing a block diagram of the system is the best way to go about fixing it. To fully grasp how an electronic system works, all one has to do is study the accompanying block diagram, which can be found in most service manuals. It's typically more useful than a full circuit diagram when a system is turned on for the first time.

Split-Half Method: When no output is detected, the circuit is cut in half and the results are first checked at the halfway point as a first step in the troubleshooting process. Locating the faulty path early or late in the game is important. As soon as the defective section is identified, it is cut in half. This will keep happening until the loss discloses a single function or part. When there is a high number of blocks in series, the half-split approach is preferable. In order to generate many types of timing pulses, an oscillator is often divided into subsets at different frequencies using decade counters. In Figure 1.10, we see one such setup.

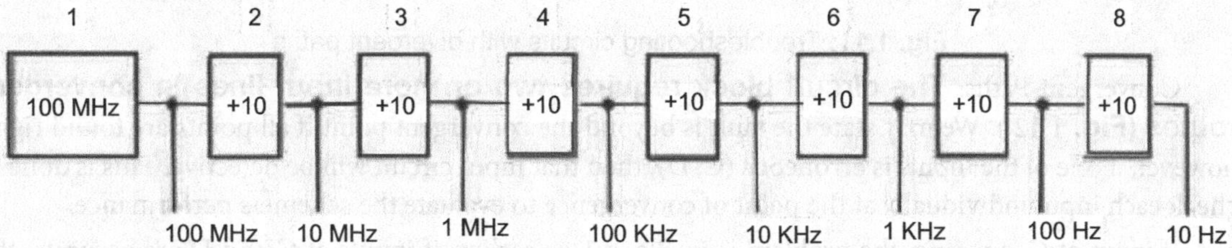

Fig. 1.10 Split-half method for isolating trouble in a circuit

Example : Let us just suppose this circuit's expected output (pulses at 100 Hz) is missing. Let's say we assume the issue is in block 4. The sequence of carrying out tests would be as follows:

- Divide the circuit in half mentally and check block 4's output. In this case, there will be no output. Thus, the issue may have occurred at any point throughout the opening period (blocks 1 to 4).
- Measure the second block's output by further dividing the circuit in half. It will be verified that 10 MHz is the right output frequencies. Therein lays the fault (blocks 3 and 4)
- Take a reading from block 3's output. The 1 MHz value will be confirmed to be accurate.Therefore, the fault lies in block 4.

This may be trying to simplify the situation by using an excessively small sample size and a low number of tests. In practise, if there are several failures in the system, additional measurements would be required to locate a defect. Unfortunately, most electronic systems do not consist of only linked blocks. In certain places of the circuit there may be feedback loops or parallel branches. Localizing errors in such cases will be more difficult, but the half-split method may still be used. The following events may be monitored closely, and a systematic fault analysis can be performed.Fig. 1.11 Illustrates a divergent route, in which the output of a single block feeds two or more more blocks. The most prevalent instance of this is the power supply circuit, which is responsible for distributing DC power to the many sub systems of machinery.

In such a system, the common feed point is checked first to see if one output is normal (let's say at A or B), after which the divergence point has been reached; however, if one output is abnormal, the common point is then checked once again.

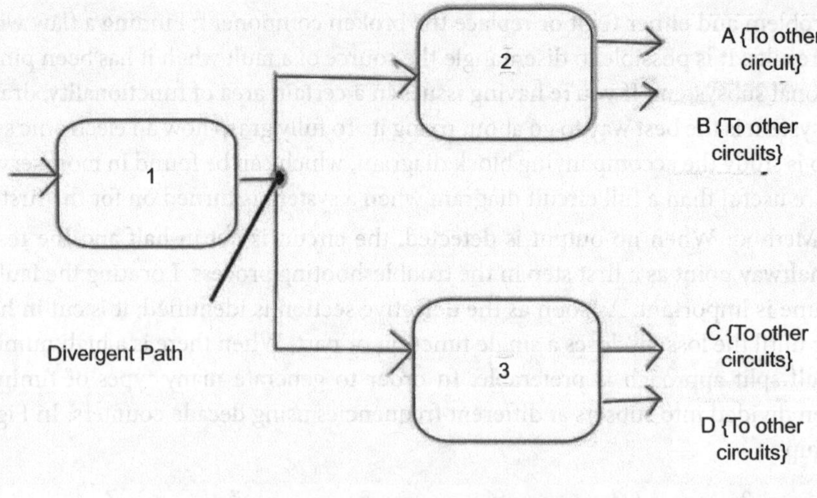

Fig. 1.11: Troubleshooting circuits with divergent paths

Convergent Paths: **The circuit block requires two or more input lines in convergent routes (Fig.** 1.12). We may state the fault is beyond the convergent point if all points are found right; however, if one of the inputs is erroneous (C, D), then that input circuit will be defective. This is done to check each input individually at the point of convergence to evaluate the scheme›s performance.

If input at C is wrong, the problem is in Block 1 or earlier; if inputs at C and D are accurate, the problem is in Unit 3, and there is no output at E.

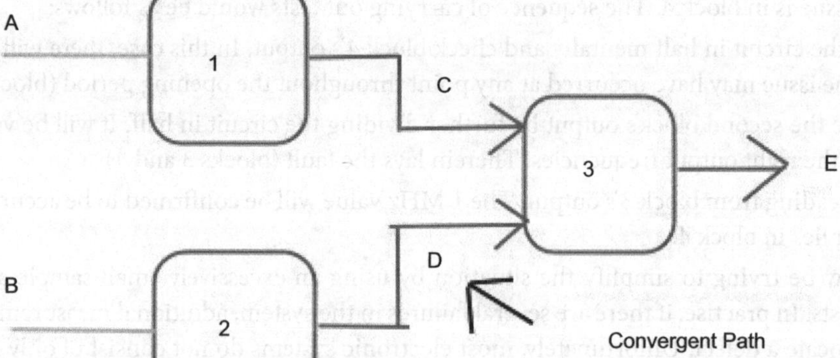

Fig. 1.12: Isolating faults in convergent paths

Feedback Paths: A feedback loop in a circuit causes more complicated issues at the point of failure. This is the procedure wherein the results of one system are used as inputs to another system, known as a feedback loop (Fig. 1.13). In a feedback loop, the final result will look bad if all of the intermediate outputs are incorrect.

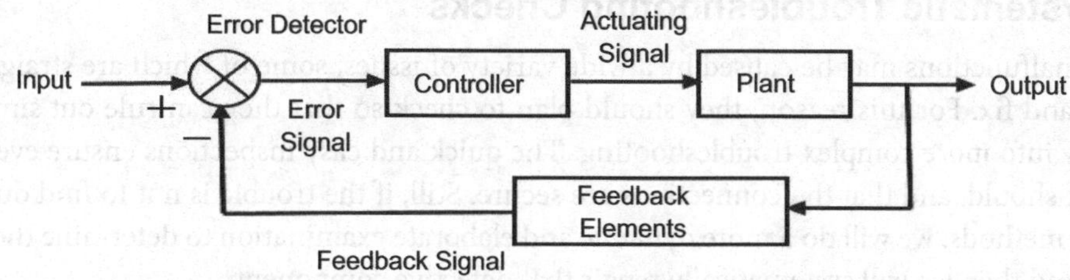

Fig. 1.13: Feedback path

Feedback paths are provided basically for the following functions:

- Output in this situation cannot exist without the feedback, which is crucial to the circuit's continued operation. Such is, the circuitry of an oscillator.
- The feedback loop is included to allow for the alteration of certain aspects of the system, such as the circuit's functioning,

If a broken system has a feedback loop, it's important to learn about the different kinds of feedback and what they're used for before trying to fix it. That way we may go *in the direction of the alleged defect.* If the feedback is modified enough, the system might be transformed into a linear data flow, breaking the feedback loop in the process. In this case, the feedback path re-joins the forward path that was the main forward path, allowing us to test each block independently of the fault signal given to the loop, as is the case in some other examples where it is preferred to modify the feedback at the point rather than completely breaking the feedback loop. If the system output is as expected, verify the feedback circuitry, and if it is not, investigate the forward route. Finding the source of the problem might be challenging because of the large range of feedback loops. For the sustained variety, the feedback is cut off from the input and replaced with a test signal that may be used to evaluate the functionality of the individual components.

Switching Paths

In a system with switchable components (Fig. 1.14), if the circuit function is failing in one position of the switch, consider swapping it to the opposite position. The switch in the building's shared circuitry should be checked if the problem persists. If the issue goes away after that, the circuitry may be at fault.

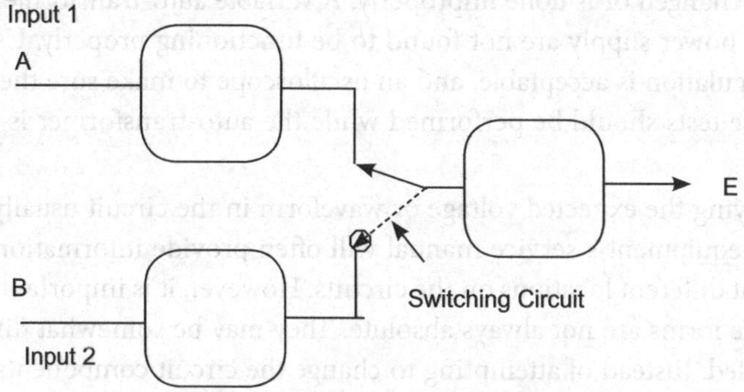

Fig. 1.14: Fault location in switching circuits

1.9.3 Systematic Troubleshooting Checks

Equipment malfunctions may be caused by a wide variety of issues, some of which are straightforward to diagnose and fix. For this reason, they should plan to check so that they can rule out simple issues before diving into more complex troubleshooting. The quick and easy inspections ensure everything is working as it should, and that the connections are secure. Still, if the trouble is not to find out with the help of these methods, we will do a more dynamic and elaborate examination to determine the defective component and then we will systematically repair that defective component.

The following steps are usually helpful in troubleshooting:

(a) Check Control Settings: Sometimes unnecessary trouble exists in equipment due to lots of controls on the front or back panel, causing wrong control panel settings for the operator. The technician should consult the operating manual for the appropriate function or operation of the control panel

(b) Check Associated Equipment: Associated equipment should operate correctly before troubleshooting the leading equipment (particularly test and measuring instruments). If an oscilloscope, for instance, isn›t operating, you should make sure the signal is connected properly and that the connections aren›t broken. Make sure the power supply is working properly.

(c) Visual Check: The problem and its location should be extensively inspected through a visual inspection. Desoldered connections, broken wires, and ruined circuit boards or components may be seen with ease by doing a simple visual inspection.

(d) Calibration: There are several issues that may be triggered by the inexperienced use of some nonprofessional calibration controllers. Therefore, you should examine the equipment›s calibration or the damaged circuit. In order to fix the issue, you may follow the manufacturer-recommended calibration method.

(e) Isolate the Troubling Circuit: Require careful note of the symptoms of the problem and conduct an in-depth analysis to determine which circuit is causing the issue. Usually the signs can tell you which circuit has the issue. If your CRO focus control is off, for instance, it might be due to a problem with the high-voltage supplies that power the cathode ray tubes. Similarly, if your electronic device›s whole electrical system is malfunctioning, the problem is likely with the power supplies. The voltages between the supply›s test points and ground may be used as an indicator of how near to specification the power supplies are operating. External voltages increase the risk that the supply is not changed or is done improperly. A variable auto-transformer may be fitted to the machinery if the power supply are not found to be functioning properly. Use a DC voltmeter to make sure the regulation is acceptable, and an oscilloscope to make sure the ripple is in the right place. All of these tests should be performed while the auto-transformer is adjusted over its full range of motion.

(f) Measuring: Verifying the expected voltage or waveform in the circuit usually identifies the faulty component. The equipment›s service manual will often provide information on normal voltages and wave forms at different locations on the circuits. However, it is important to keep in mind that voltages and wave forms are not always absolute. They may be somewhat different depending on the machinery used. Instead of attempting to change the circuit components to reach the correct voltage levels and waveform patterns, these measurements and observations should be utilized as a reference for identifying a malfunction. Still, every major deviation has to be investigated and fixed for the machinery to work correctly.

(g) Individual Components: Suspect components should undergo thorough testing. In order to ensure that components have been properly soldered, it is recommended to first remove one end. Therefore, the power must be switched off before removing or changing semiconductor components, isolating the measurements from the influences of the surrounding circuitry.

Some helpful troubleshooting tips are given below:

1. The modular nature of certain pieces of equipment makes it easier than others to swap in a working module. The aim now is to slow down the issue, which this will achieve.
2. Many issues have simple remedies, but since they are not adopted right once, the issue becomes complicated and requires careful assessment. There are mostly mechanical structure-related problems. Technician knowledge and keen observation suggest that these issues may be remedied with the use of appropriate precision hand tools, light oil, grease degrease, and mild alcohol.
3. Make sure the issue is confirmed before attempting any fixes. In order to solve the issue, it's important to get as much information as possible from the equipment user. 'Did the issue come and go before eventually remaining awful for good?' is an example of a question that might serve as a useful troubleshooting hint.

1.9.4 Rules of Thumb in Troubleshooting

Followings are the rules of Thumb in Troubleshooting

(a) Power supply failure is a common cause of unit death or failure across several functions. Most of the time, they are simple to identify and fix.

(b) In most cases, faulty connections are to blame for problems that suddenly appear quickly. Possible causes include overheated solder junctions and dirty, unseated connections on the inside or outside of the device.

(c) Problems that develop with time. They are typically caused by dried-out electrolytic capacitors and lessen or go away when the equipment warms up. While failure can be easily diagnosed when they leak, this is not the norm. Therefore, another ESR test is required.

(d) When something catastrophic fails, it usually leaves behind charred, burned, shattered, burst, or melted parts. For a quick identification, just use your eyes and nose to look around for any telltale signs.

(e) Some parts, such as the fly-back transformer, yoke, or other magnetic components, may also make a buzzing sound at various intervals. It is important to differentiate between the two sorts of noises.

(f) Electronic devices such as televisions, video recorders, computer displays, etc., are prone to a wide variety of common issues, and as a result, several groups have established a database detailing these issues. Case histories and real-world repair experiences form the basis of the data they present. Although some businesses make their data freely available online, most do not. If that's the case, you can do a thorough analysis of the problem.

1.10 APPROACHING COMPONENTS FOR TESTS

In order to facilitate testing, most manufacturers provide test points strategically placed throughout the PCB. Specific DC and AC voltages, as well as the waveform, characterize these locations. To see a test point on a PCB, refer to Figure 1.15. A test prod may be connected to the pin, which is typically mounted vertically.

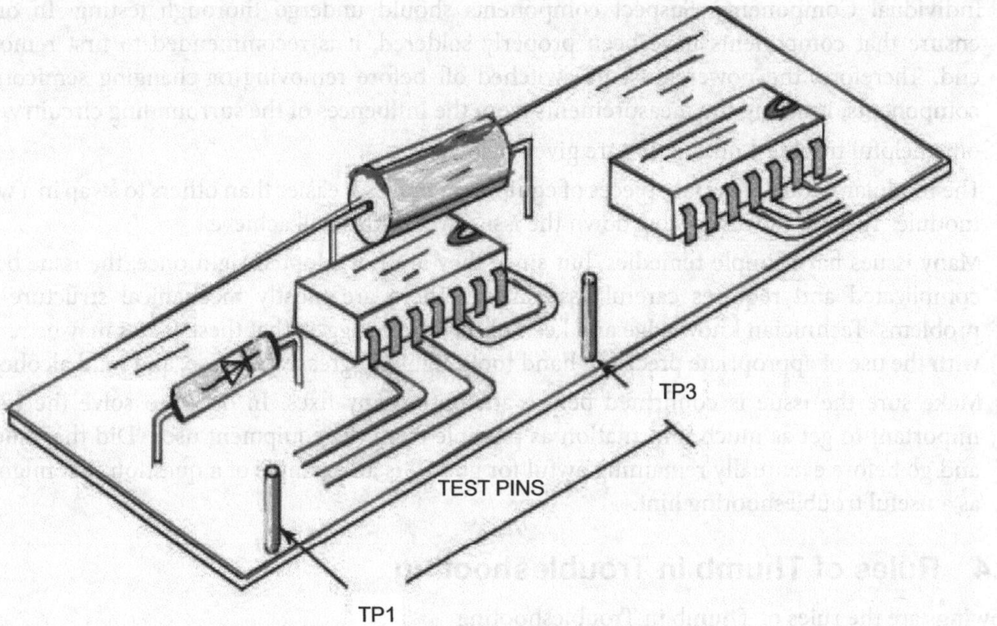

Fig. 1.15: Printed circuit board showing test point

The circuit may be measured at numerous locations if no designated test sites are supplied, as shown in several parts becoming close together.In that case, proceed as follows:

(a) Prod the legs beneath the casing to see whether they are connected to the transistor.

(b) Finding a connected component is necessary for reading the signal from a trace on a circuit board(Fig. 1.16).

(c) Connect your test lead to the trace by attaching it into the leg of the component.

know more about troubleshooting method

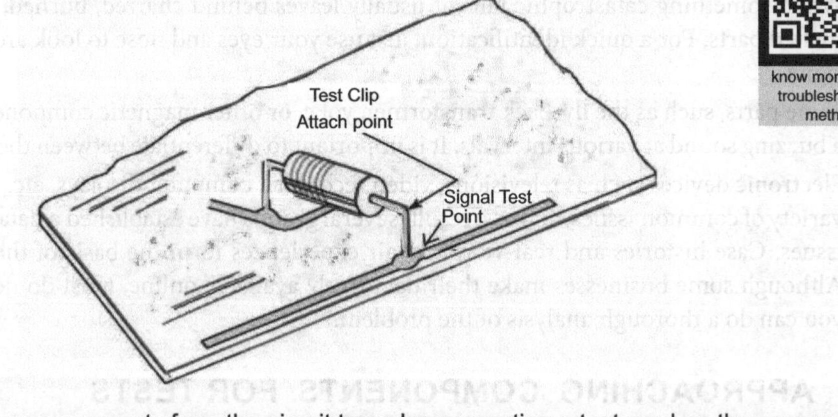

Fig. 1.16: Taking measurements from the circuit trace by connecting a test prod on the component

Using an IC test clip (Fig. 1.17) to make connections to the integrated circuits is more convenient since no more than one conductor has to be handled at a time. A short circuit may be simply made otherwise. Since digital circuits are often packed closely together on a board, a light touch is required while probing them.

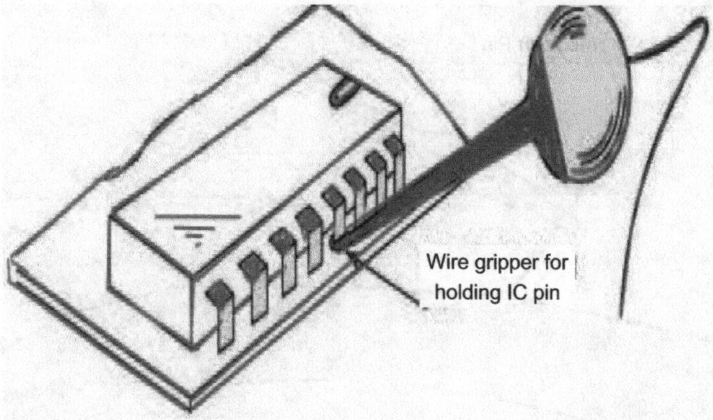

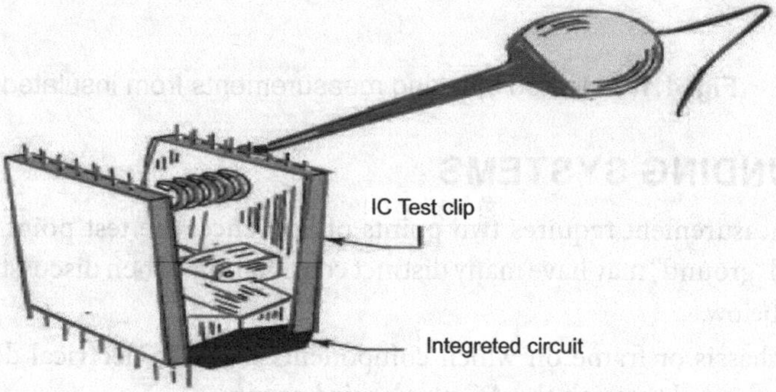

Fig. 1.17: Use of test clip for taking measurement on IC

One of the best places to take readings is on a flexible flat wire that has connections. While the connection pins themselves are often shielded, the wires behind them may be monitored.(Fig. 1.18).

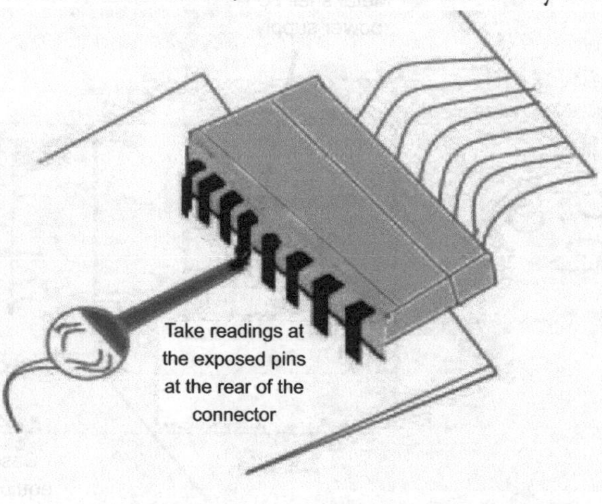

Fig. 1.18: Taking test readings from a connector

As may be seen in Fig. 1.19, a reading from an insulated wire can be obtained by puncturing the insulation. Needle the insulation and attach the test prod carefully. After the test is complete, wrap the damaged wire with insulating tape.

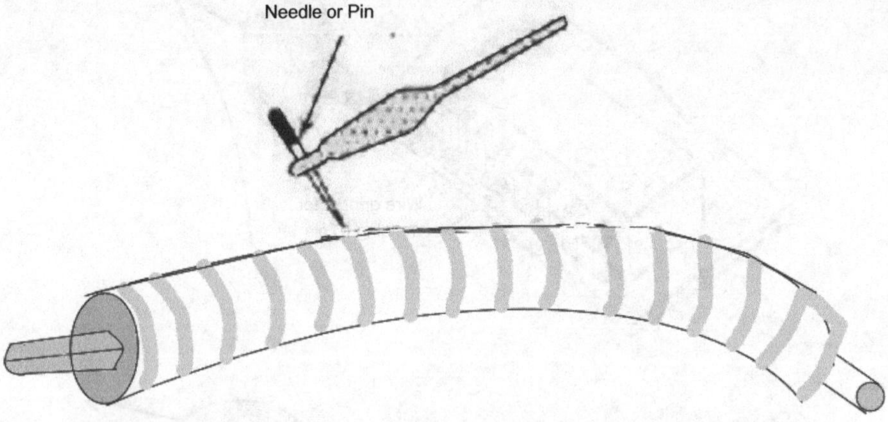

Fig. 1.19: Method of taking measurements from insulated wire

1.11 GROUNDING SYSTEMS

Each electrical measurement requires two points of reference: the test point being measured and the ground. The word "ground" may have many distinct connotations when discussing electronic component, as will be shown below.:

(a) The metal chassis or frame on which components in older electrical devices were attached was normally grounded through the AC plug's third prong.

(b) Grounding traces on circuit boards, ground leads on motors, and ground leads on solenoids are all examples of chassis grounding, as seen in Figure 1.20.

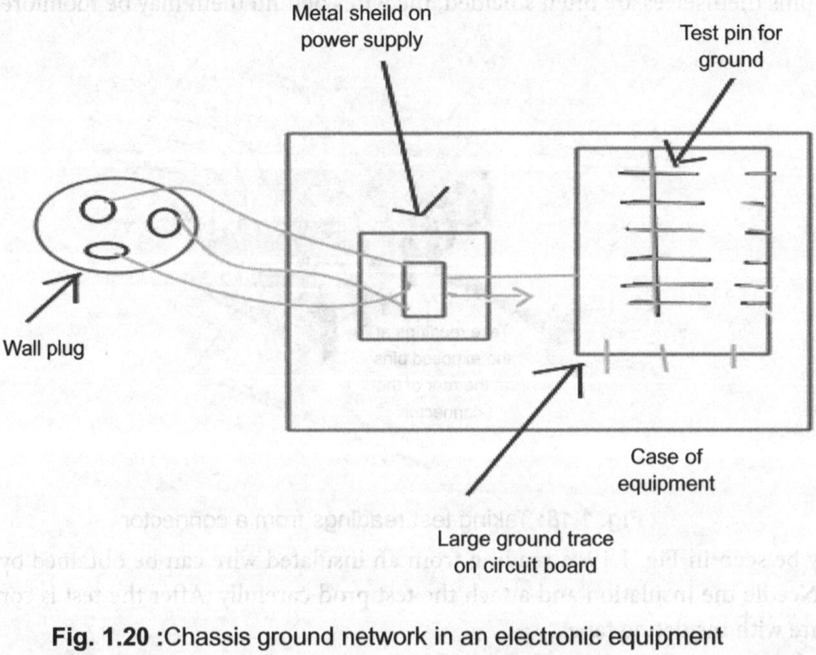

Fig. 1.20 :Chassis ground network in an electronic equipment

(c) The chassis ground traces on a PCB are typically broad and located towards the board's edge. From there, they connect to the rest of the circuit. Typically, this ground has its own test pin on the circuit board.

(d) 'Logic ground,' which is separate from the chassis ground, is often supplied for the digital circuits (TTL) in most pieces of equipment. Fig.1.21 depicts such a configuration.

(e) The power supplied to the TTL ICs is kept pure by this grounding mechanism, which shields it from transients and electrical noise. The logic ground in this system will have a somewhat different potential than the chassis ground.

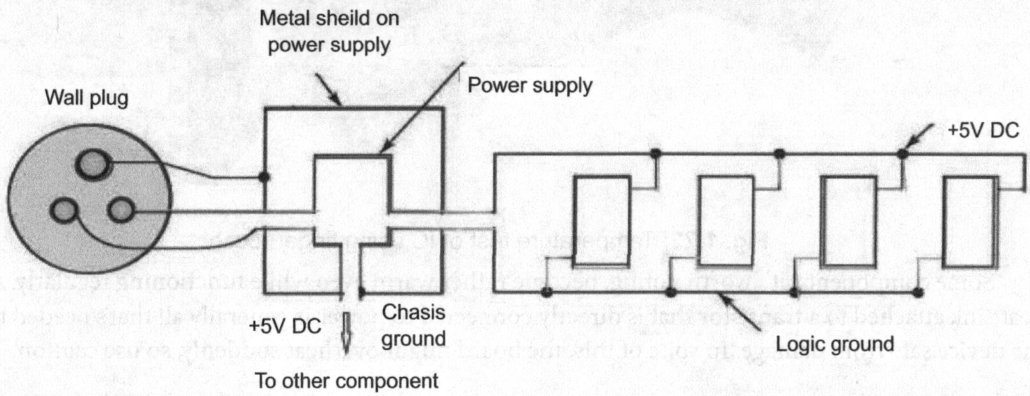

Fig. 1.21: Chassis ground and logic ground in a digital electronic equipment

If 'signal' ground is also present, then a third sort of ground may be found. This separate ground is used as a reference for the signals. Such a foundation may be seen in the connection between a computer and a disc drive. Given the variety of equipment grounds, it's important to establish a connection to ground as near to the component being tested as feasible. It will clarify the differences between the three soil types.

1.12 TEMPERATURE-SENSITIVE INTERMITTENT PROBLEMS

Identification of that troubles which do not exist is more difficult task for any technician. The rarest kind of problem is the one that disappears as soon as you begin seeking for a solution. In many cases, you may induce such state by increasing the temperature of one of the boards. In most cases, the heat from a table lamp or a household hand-held hair dryer will be sufficient, but in others, you may soon overheat the board.

You can generally get the task done very fast by combining the heating act with a cooling act when you notice the problem is not continuous, occurring sometimes heat create or sometimes not on to the full region. Spray the heated board and its various components with circuit coolant from a pressurized container. The problem will appear as soon as the faulty part is cooled. When everyone else has been chilled, it becomes irrelevant. The burst of coolant need to be quick and localized only on the part that is thought to be malfunctioning. If you believe you've found the source of the problem, keep in mind that you could really just be near.

The bad Integrated Circuits often have a warm to hot sensation. If you want to quickly identify the problematic IC, a "touch test" may be performed with your finger. The back of your finger is ideal for this examination, as seen in Fig. 1.22 You may feel heat here before it reaches the tip of your finger, which is uncommon. Technicians often use digital thermometers to check the temperature of a component they suspect of being defective.

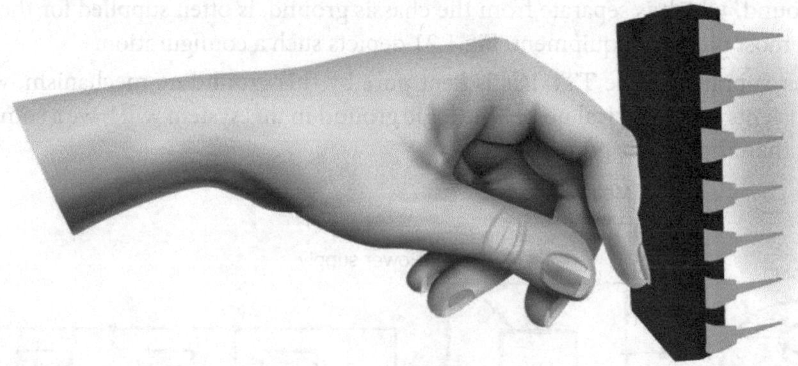

Fig. 1.22: Temperature test of IC using finger touch

Some components, it's worth noting, become rather warm even while functioning regularly. A large heat sink attached to a transistor that is directly connected to power is generally all that's needed to keep the device safe from damage. In spite of this, the board might overheat suddenly, so use caution.

1.13 CORRECTIVE ACTION

After determining the issue and pinpointing the faulty part, repairs will need the use of specialize tools and methods. Below you'll find an outline of the procedures involved.:

1.13.1 Arranging Replacement Parts

Most of the time, you can get new electrical and mechanical components straight from the manufacturer of the device you need to fix. However, many of the common electrical components may be purchased from suppliers, which can speed up the procurement process significantly. Careful consideration of the value, tolerance, rating, and description of the component needed should precede either its purchase or a request to the manufacturer for replacement components.

Care should be used when choosing a component since a different composition might negatively influence equipment performance; it is worth keeping in mind that the physical size and shape of a component can occasionally affect its performance in the equipment.

The equipment may include non-standard electronic components in addition to the usual suspects. Equipment makers create or source such parts to achieve predetermined benchmarks. The mechanical components are custom constructed and can only be supplied from the machine's manufacturer.

Equipment vendors need the following data in order to make early orders for replacement parts:

(a) Name of the equipment;
(b) Equipment model or type
(c) Equipment serial number;
(d) Specifics about the required component, such as its brand and part number.

(e) This is the part number from the manufacturer's service manual that identifies this specific component. Upgraded parts become available from manufacturers, and occasionally this requires modifying existing equipment. They often send updates to the service manual in the form of "addenda" that detail such changes. If you find that a piece of equipment doesn't match the components on the schematic, check for an accompanying "modifications document." If there is any confusion, the maker may be contacted.

1.13.2 Component Replacement

The device's power supply must be cut off before any attempted component replacement may be made. The service manuals frequently provide understating view drawings connected with the mechanical parts, and they may be straightforward in the removal or disassembly of individual components of sub-assemblies. Avoiding component replacement until absolutely essential is a good practise. In the alternate, it may cause harm to the circuit board and any nearby components.

Some damage to a circuit board may be irreparable, in which case the whole thing, including any components that were soldered on, will need to be replaced. Boards may be acquired from the producer of the device.

Replace semiconductor components only if testing reveals they are broken. Replacing semiconductors without cause might throw off equipment performance and calibration. Always replace bulbs in their original sockets after removing them for maintenance. Any semiconductor components that need replacing must be identical or functionally equivalent to the originals. Some plastic-cased transistors deviate from the norm because they employ lead arrangements that are more often seen in metal-cased transistors.

Because of the vulnerability of the pins, removal of the integrated circuits (dual-in-line package, 14- and 16-pin varieties) requires an extraction tool. You may pull carefully and evenly on both ends of the gadget if removing equipment is not conveniently accessible. Be careful not to harm the pins on the socket by removing the integrated circuit from one end before the other.

After replacing a power transistor, make sure there is no short between the collector and ground. These transistors are often positioned on the absorber plate. If the equipment's switches are faulty, it's typically not possible to fix them; instead, the whole piece of machinery must be changed out.

Unless it is battery-operated, almost every piece of electronic equipment needs a power transformer. Power transformers must be replaced with identical models in the event of a failure. Mark the cables with the terminal numbers when you remove the transformer to facilitate the repair. Check the performance of the complete equipment. after the transformer has been replaced.

1.13.3 Performance Check

When an electrical item is changed, it's important to test the circuit to ensure that the new part didn't throw off the calibration or performance. When determining whether or not this portion has been worked on or the power transformer has been changed, it is necessary to inspect all of the equipment, since power supplies are utilized to run all of the circuits. If the tolerance specified in each 'CHECK' section is not fulfilled, the relevant component of the circuitry must be adjusted so that other sections are not forced to be readjusted.Terms like these are often seen in service manuals and relate to doing a performance check :

(a) Check: Signifies that a performance requirement check has been completed as a result of the instruction. An adjustment or repair may be necessary if a monitored parameter is outside of the specified range.

(b) Adjust: Specifies which tweak should be made and how it should affect the system. No alterations are to be performed until a 'Check' command has been issued and confirmed as appropriate.

(c) Interaction: This instruction emphasis the interaction between the previously mentioned circuit modification and the other possible circuit adjustments. Indications of the interaction's nature and mention of the processes it affects are provided.

1.13.4 Replacement of Circuit Boards

The service/instruction manual provided with the equipment usually offers enough information to let a competent electronic technician diagnose and fix any circuits that are causing the device to fail. If a fault is found on a single or several boards and cannot be fixed locally, the boards might be sent back to the manufacturer to be replaced or repaired. For the most time and money saving when replacing any circuit board, many manufacturers suggest ordering the exchange board. Its cost is often much lower than that of a brand new board.

The following is often required by the manufacturer when seeking an exchange board

(a) Equipment Description: Information about the equipment (model, serial number, and manufacturer's name). All of this data is typically shown on the front and back sides of the unit.

(b) Part Number of the Board: The components list included in the manual will often provide this kind of data. The number on the PCB is not the actual component number.

(c) Purchase Reference: This information is useful for determining the unit's warranty status and the appropriate method of payment.

Know more about PCB board

1.14 SITUATIONS WHEN REPAIR NOT ATTEMPTING

The proper appraisal of monetary, emotional, or other value plays an important part in determining whether or not it is worthwhile to repair the equipment. The equipment is not cost-effective to repair, however, under the following circumstances :

1. Damage of a catastrophic kind caused by smoke, fire, or water.
2. Physical harm, particularly to the mechanical components, necessitating total replacement if they are to be used again.
3. Machines where previous maintenance efforts may have introduced a plethora of new, unexpected circumstances.
4. Machinery that has been shown to have design or production flaws, for which a fix is not yet available.

5. Cellular phones, pagers, corded phones, computer motherboards, disc drives, etc. might be challenging to repair without the proper paperwork, test equipment, network infrastructure, or replacement parts. A dead power supply, a damaged connection, or a broken trace on the PCB are all easily fixable causes of malfunction.
6. In the event that you are really clueless about the machinery, it is best to let a skilled specialist examine it. It's not only possible that you may be hurt if you do this, but you also risk doing irreparable damage to the circuitry.
7. Avoid fixing machinery unless you have the proper diagnostic tools and equipment. Removing a component off a multi layer PCB without soldering equipment will result in a catastrophic mess.

1.15 GENERAL GUIDELINES

- Conduct regular inspections your test equipment.
- Never assume that brand-new parts are guaranteed to function properly.
- Avoid assuming there will be only one accident by doing the following: It is not uncommon for many failure mechanisms to be linked to a single symptom.
- It is not necessary to assume a causal relationship between events that occur close in time. They may all be the result of the same thing, or they could have nothing to do with one another at all.
- After working on the same problem for a while, it's easy to lose sight of important details. Give it to someone else to look at, or at least ask for their assistance. It's time to take a break. You will feel wonderful

UNIT SUMMARY

- Drawing of any electrical circuit is a graphical representation of connections and components. It can be a wiring diagram, electrical diagram, elementary diagram, block representation, for designing, building, and troubleshooting purposes
- Opening or dismantling and reassembling equipment is a common troubleshooting technique. However, disassembly and reassembly calls for careful attention, especially if getting inside the apparatus seems hard.
- The challenging part of any equipment repair is opening the equipment non-destructively. There are many elements to secure electronic equipment, including screw,hidden screw,snaps etc
- After equipment repair, all parts should be reassembled correctly together without any external force
- A catastrophic failure occurs when there is an abrupt or total change in the functioning of the device. It might be due to a short circuit or an open circuit that cannot be repaired.
- Failures in machinery can occur for a number of causes. As a whole, they may be summed up as things like sloppy planning, botched manufacturing, improper storage, and unsafe working conditions.
- The failure modes of electrical devices can range from straightforward mechanical failures to more complicated electronic circuitry failures.

- The technician's skill and familiarity with the machine will determine how quickly and easily the problem can be located. Obviously, when troubleshooting a system, nothing can replace expertise with the equipment.
- Common troubleshooting resources include (a)service and maintenance manuals and instructions; (b)test and measuring instruments; and (c)specialized tools (instruments, mechanical tools)
- The three primary tools can be used to detect and repair malfunctions: multi-meter oscilloscope,function generator
- It is crucial to do preliminary examinations prior to really repairing broken equipment. For your own good, these inspections are essential, and they also speed up your arrival at the scene of the crisis.
- It is crucial to do preliminary examinations prior to really repairing broken equipment. For your own good, these inspections are essential, and they also speed up your arrival at the scene of the crisis.
- After determining the source of the problem and identifying the faulty part, repair work requires specialize skills.
- Most equipment's original manufacturer is the best source for replacement electrical and mechanical parts.
- Temperature-sensitive intermittent difficulties emerge when a device fails because of incompatibility with the ambient temperature.
- Bad ICs usually have a warm or hot sensation. A defective IC can be discovered with a fast "touch test," which can be performed with your finger. Making the test using the back of your finger is preferred.
- Modern machinery's grounding traces on circuit boards, ground leads on motors, and grounding solenoids are all examples of what are meant by the term "chassis ground."
- To ensure that the power supplied to the TTL ICs is clean and free of transients and electrical noise, a grounding mechanism has been implemented. The logic ground in this system will have a somewhat different potential than the chassis ground.
- Most manufacturers also include test points in easy-to-access areas on the PCB. Specific DC and AC voltages, as well as the waveform, characterize these locations.
- Components often suffer severe damage, including burning, scorching, cracking, exploding, or melting, after catastrophic failures. For a quick identification, just use your eyes and nose to look around for any telltale signs.
- If you change an electrical item, make sure to double-check the circuit's accuracy and functionality.
- It is possible to send back a faulty board to the maker for a replacement or fix if the problem is isolated to that board.
- To request an exchange board from the manufacturer, you will need to provide the following details:
- Board Description, Purchase Reference, and Part Number
- Damage of a severe kind caused by water, fire, or smoke, If the equipment has sustained severe physical damage, it is not worth attempting to restore it.
- Don't think there's only one possible failure. It is not uncommon for many failure mechanisms to be linked to a single symptom.

- Unless it is battery-powered, virtually every piece of electronic equipment requires a power transformer. Power transformers must be replaced with identical models in the event of a failure.
- If a technician follows the procedures given in this chapter, he or she will be able to diagnose electronic equipment in an organize fashion and perform repairs and maintenance in a timely fashion.

EXERCISES

Multiple Choice Questions

1. In electronic diagrams, which items are not necessary to be shown if they are not asking for clarification:
   (a) Transmission diagrams (b) Operating relays
   (c) Closed relays (d) Terminal circles
2. A decreasing succession of sine waves is known as a
   (a) Ringing (b) Slew
   (c) Overshooting (d) Undershooting
3. What comparability between input and output signals can be tested using the best equipment?
   (a) An Oscilloscope (b) A Logic Probe
   (c) A Spectrum Analyzer (d) A Multi-trace Oscilloscope
4. What is the next step after discovering a faulty gate within an IC?
   (a) Repair The Gate
   (b) Resolder The Tracks
   (c) Replace The Ic Involved
   (d) Recheck The Power Source
5. The use of a multi-meter with digital circuits allows the measurement of:
   (a) Pulse Width (b) Voltage Or Resistance
   (c) Current (d) Pulse Trains
6. In order to predict the usable service life of an electronic equipment what is basic requirements.
   (a) The mean random failure rate, (b) The mean wear-out failure rate.
   (c) Both (a) and (b) (d) others
7. The failure of an equipment not associate with
   (a) Inability to perform function (b) wrong circuit diagram
   (c) short-circuit in components (d) location of equipment
8. Reason of Production De ficie ncie s is
   (a) Lack of inward inspection of goods (b) Not suitable storage methods
   (c) Poor motivation, lack of skill and sense (d) All of above
9. Fault-Finding Aids means
   (a) Fault localization and its subsequent repair
   (b) Hide faulty component

(c) First -aid treatment
(d) Location of equipment

10. Following is not measuring and testing instrument
(a) Multimeter (b) Ohmmeter
(c) Oscilloscope (d) Power supply Cord

11. What is Name of Technique in which the circuit is split into half and the output is checked at the half- way point
(a) Split-Half Method (b) Wrong Division Method
(c) Distributive Technique (d) None of these

12. Which path have two or more input lines feed a circuit block
(a) Divergent Path (b) Convergent Path
(c) Switching Path (d) Feedback Path

13. Which is not helpful step for troubleshooting
(a) Check Control Settings (b) Calibration
(c) Isolate the Troubling Circuit (d) Equipment Brand and purchase Date

14. In many instances, the bad ICs will feel warm or hot to the touch. Find the faulty IC with a simple "touch test" using your finger. You should check with the pad of your finger instead of the tip Because
(a) This part of the finger is actually more sensitive to heat
(b) Digital thermometers to not giving correct answer
(c) ICs outer layer is not sensitive
(d) That technique increases equipment health

15. In which Situations repairs should not be attempted
(a) Serious damage due to water, fire or smoke.
(b) Serious physical damage, especially to the mechanical parts, unless they are completely replaced.
(c) Machines where previous fixes may have introduced other issues that have yet to be identified.
(d) All of these

ANSWERS															
1.	(d)	2.	(a)	3.	(d)	4.	(c)	5.	(b)	6.	(c)	7.	(d)	8.	(d)
9.	(a)	10.	(d)	11.	(a)	12.	(b)	13.	(d)	14.	(a)	15.	(d)		

Short and Long Answer Type Question

Category 1

1. Define common type fault in capacitor
2. Why power supply consideration in instruments need to be taken care of?
3. Demonstrate the fact that "Physical inspection of an equipment is necessary before its maintenance,"
4. State any two type of failure in electronic equipment

5. short note on Split half method
6. Explain feedback path circuit analysis
7. Differentiate between preventative and corrective maintenance explain Applying Procedure on localize a fault in old equipment
8. Analyze all the safety parameter can we taken While troubleshooting on electronic equipment
9. What do you understand by the term troubleshooting,repair,maintenance and servicing
10. Evaluate difference between block diagram and circuit diagram.

Category 2

1. Analyze the guidelines for drawing diagrams the main features of these guidelines
2. Summarize the reasons of Production deficiencies
3. Define Nature of fault .
4. Describe the fault location procedure in details
5. Discuss the most commonly needed fault finding aids with the help of examples.
6. List the testing and measuring electronic instrument by which Faults can be located and rectified
7. Explain all the Troubleshooting techniques in details.
8. Summarize and remember the basics points for an effective system of troubleshooting
9. Differentiate among Feedback path, Switching path, convergent path,Divergent path.

REFERENCES AND SUGGESTED READINGS

1. Cooper, W.D. (1984), Electronic Instrumentation and Measurement Techniques, Prentice Hall of India Private Ltd., New Delhi. ISBN 0-87692-060-1
2. Gottlieb, I.M. (1985), Power Supplies; Switching Regulators, Invertors and Convertors, PBP Publications, Delhi.Harper, C.A. (1977),
3. Handbook of Components for Electronics, McGraw-Hill Co., USA, ISBN 0-07-026682-4.
4. Khandpur, R.S. (1996), Handbook of Analytical Instruments, Tata McGraw- Hill Publishing Co. Ltd., New Delhi. ISBN 0-07-460186-5
5. Loveday, G (1989), Electronic Testing and Fault Diagnosis, Longman (Scientific and Technical), England. ISBN -582-03865-0
6. Malvino, A.P. (1989), Electronic Principles, 4th ed. McGraw-Hill, U.S.A. ISBN 0-07099479-X.
7. Robinson, V (1974), Handbook of Electronic Instrumentation Testing and Troubleshooting, Reston Publishing Co., U.S.A. ISBN 0-87909-327-7.
8. Scarlett, J.A. (1984), An Introduction to Printed Circuit Board Technology, Electrochemical Publications Limited. U.K. ISBN 0-901150-16-9.
9. Sharma, D.D. (2000), Total Quality Management, Sultan Chand & Sons, New Delhi. ISBN 81-7014-623-2.
10. Wassink, R.J.K. (1984), Soldering in Electronics, Electro chemical Publishers Limited, Britain. ISBN 0-901150-142.

1. https://nptel.ac.in/courses/108101167
2. https://archive.nptel.ac.in/courses/112/105/112105232/
3. https://www.dv-group.com/en/maintenance/electronic-maintenance/
4. https://blog.fhyzics.net/sop/sop-manual-for-other-electronics-and-precision-repair-and-maintenance

KNOW MORE

When any troubleshooting problems s begin, usually comply with the necessary precautions to shield records on a computer. a few upkeep, which include replacing a hard force or reinstalling an operating machine, may position the records at the computer at danger.A statistics backup is a duplicate of the data on a computer hard drive stored to another storage device or to cloud garage. Cloud storage is online storage that is accessed through the internet. In an organization, backups may be done on a daily, weekly, or monthly foundation.

□□

2 Passive Component and Their Testing

UNIT SPECIFICS

Through this unit we have discussed the following aspects:

- Understanding the passive component like resistor,capacitor,inductor
- Understanding the potentiometer LDR and thermistor
- Testings of passive component
- Measurements of passive components.

The practical applications of the topics are discussed for generating further curiosity and creativity as well as improving problem solving capacity.

Besides giving a large number of multiple choice questions as well as questions of short and long answer types marked in two categories following lower and higher order of Bloom's taxonomy, assignments through a number of numerical problems, a list of references and suggested readings are given in the unit so that one can go through them for practice. It is important to note that for getting more information on various topics of interest some QR codes have been provided in different sections which can be scanned for relevant supportive knowledge.

After the related practical, based on the content, there is a "Know More" section. This section has been judiciously designed so that the supplementary information provided in this part becomes beneficial for the users of the book. This section mainly highlights the initial activity, examples of some interesting facts, analogy, history of the development of the subject focusing the salient observations and finding, timelines starting from the development of the concerned topics up to the recent time, applications of the subject matter for our day-to-day real life or/and industrial applications on variety of aspects, case study related to environmental, sustainability, social and ethical issues whichever applicable, and finally inquisitiveness and curiosity topics of the unit.

RATIONALE

This unit on "Passive components and their testing" helps students to understand the knowledge about the component like resistor capacitor and inductor etc. This unit explains the variety of components and their limitations and complete knowledge of components is an essential part of troubleshooting in electronic circuitry. And it would reduce the chances of failure . This chapter develop correct methodology of component testing so at the time of fault these components ensure top-quality performance

PRE-REQUISITES

Physics: *Semiconductor Electronic: Material, Devices and Simple Circuits (Class XII)*

Moving Charges and Magnetism (Class XII)

UNIT OUTCOMES

List of outcomes of this unit is as follows:

U2-O1: Describe the concept Resistors, Capacitors, Inductors Failures in fixed resistors

U2-O2: Describe variable resistors, variable resistors as potentiometer,

U2-O3: Explain testing of potentiometer, servicing potentiometer, LDRs and Thermistor

U2-O4: Explain types of capacitors and their performance,

U2-O5: Describe the inductance and capacitance measurement.

Unit-2 Outcomes	EXPECTED MAPPING WITH COURSE OUTCOMES (1 - Weak Correlation; 2 - Medium Correlation; 3 - Strong Correlation)					
	CO-1	CO-2	CO-3	CO-4	CO-5	CO-6
U1-O1	1	1	3	2	3	1
U1-O2	2	1	4	1	1	2
U1-O3	3	2	3	2	3	1
U1-O4	1	2	4	2	2	2
U1-O5	2	3	3	1	3	1

2.1 INTRODUCTION

Electronic circuits strongly depends upon on passive components; without them, operation would fail or the circuit would become unstable. No energy is produced by passive components , yet they may store or release energy. Resistors, inductors, capacitors, and transformers are all examples of passive devices that are essential to the construction of any electronic circuit. Due to their inherently non-unity gain, passive components in a circuit act as attenuator s rather than amplifiers or direction finders. The lack of power gain in passive components means that they can't be used to multiply a signal by more than one, but they may be used to operate complicated circuits or signals, induce a phase shift to the signal, or offer feedback.

Although they may seem like they would save power in a circuit, passive devices actually use more energy than they produce . Although some devices have a negative coefficient, passive device like resistance and capacitance always have positive values (value > 0) in 'ohm' and 'farads' respectively Passive parts, such as electrolytic capacitors, may be connected across a circuit in either direction unless specifically prohibited. The polarity of the voltage between them is established by the usual current flow from the positive to the negative terminal. These are most basic passive electrical components: resistance, capacitance, and inductance. Passive devices are sometimes referred to as "electrical elements" in the settings of both electrical circuit theory and circuit analysis. Precision measurements are necessary for passive components like resistors, capacitors, and inductors.

2.2 PASSIVE COMPONENT- RESISTOR

Resistors are the important passive components. Resistors are a significant element of electrical circuits because they act as a barrier to current flow, allowing the user to regulate the amount of current flowing through the circuit.

Types of resistors and their applications.

The diverse roles of the resistor include voltage splitting, heat-generating, LED powering, gain adjusting, matching & loading circuits, time limitations fixing. There are numerous kinds of resistors available according on the construction, application, tolerance, properties, and power dissipation. Resistors are available in numerous forms as well as sizes which have varying qualities dependent on the construction and production. While designing a circuit, these will help in knowing the benefits as well as unique functions for every resistor. Generally, resistors are separated into two categories such linear resistors and nonlinear resistors.

2.2.1 Linear Resistors

Once temperature and voltage are applied, these resistors' values may be changed. When current is flows across most linear resistor configurations, a voltage drop results. Fixed resistors and variable resistors are two types of linear resistors.

Fixed Resistors

The fixed kind of resistor is the one that is used the most. These resistors have a fixed resistance value, as their name suggests. The diversity of materials used in a fixed resistor may affect resistance characteristics including noise, cost, and tolerance.

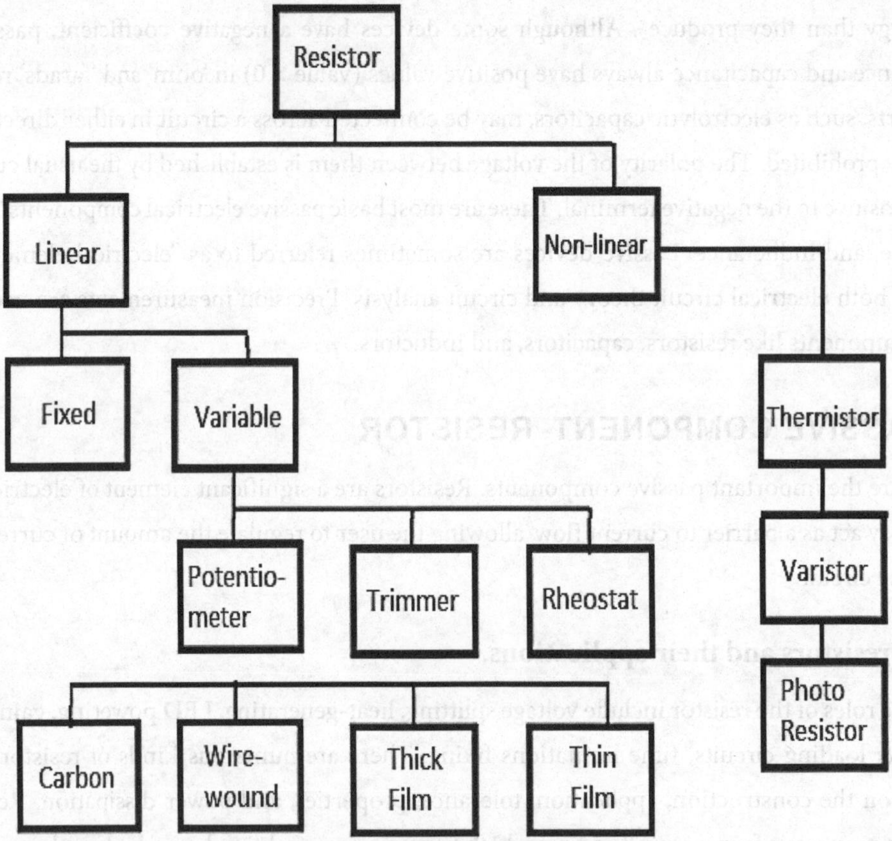

Fig. 2.1: Types of Resistor

Fixed resistors are available in different types like the following.

1. Wire Wound Resistors
2. Carbon Composition Resistors
3. Cermet Oxide Resistors
4. Metal Oxide Resistors
5. Thin Film Resistors
6. Thick Film Resistors
7. Fusible Resistors
8. Carbon Film Resistors
9. Metal Film Resistors

Know more about resistance measurement

1. Wire Wound Resistors

A wire-wound resistor is used to restrict the flow of current inside a circuit. This resistor may be made by winding conductive wire around a non-conductive core. The resistance value is determined by the conductive wire, which is manufactured with a variety of diameter and metals. The core is commonly constructed with ceramic or fiberglass,They are suited for high accuracy and power applications and not suitable for applications with frequencies greater than 50kHz.

and the resistance wire is made up of nickel-chromium alloy Wire Wound Resistors typically have minimal noise and stability in relation to temperature changes. The carbon composition resistors having different resistance values ranging from 1Ω to 22 MΩ. Fuse and circuit breaker applications, as well as other high-power industrial usage, are typical application to use these type resistor.

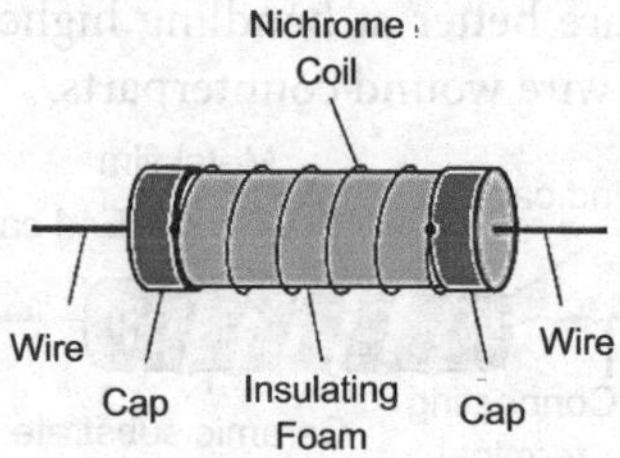

Fig. 2.2: Wire wound Resistor

2. **Carbon Composition Resistors**

Carbon composition resistors are a type fixed resistor used to limit the amount of current that may pass through them. Even though these resistors date back decades, they are seldom utilized now because of their low stability and expensive price. Carbon resistor or carbon composition are other names for carbon composition resistor. The resistors in this set have resistance ratings between 1 ohm and 22 Mega ohms Welding controls and power supply are two common uses for resistors with a carbon content.

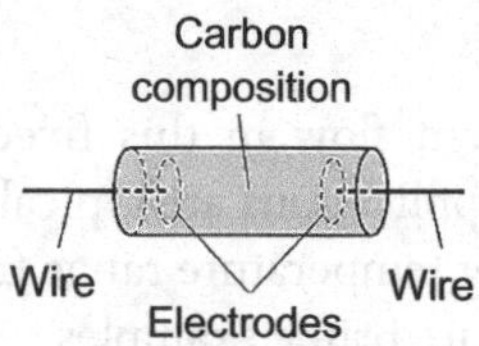

Fig. 2.3: Carbon Composition Resistor

3. **Cermet Oxide Resistors**

Ceramic insulating material and a metal alloy film or layer or carbon are arranged inside a cermet or ceramic metal to form the resistor's internal area. These resistors are built with the pins or leads below the resistors for easier placement on printed circuit boards, and they may be either rectangular or square in form. Because their values are not affected by changes in temperature, cermet oxide resistors maintain reliability even while operating in extreme conditions.

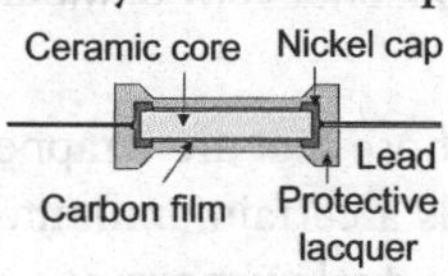

Fig. 2.4: Cermet Oxide Resistors

4. **Metal Oxide Resistors**

Such resistors may be purchased in both and fixed configurations. A ceramic rod coated with a thin metal oxide coating, such tin oxide, may be used in its construction. Metal oxide varistors (MOV) are not to be confused with these resistors. MOVs are often made of silicon carbide or zinc oxide. For this resistor's design, an oxidized coating of Tin Chloride on a substrate is all that's required. These resistors can withstand extremely high temperatures while still providing a wide variety of resistance values. In addition, high-voltage applications may benefit from the very low operating noise level. In most cases, metal film resistors are constructed from Nichrome or tantalum nitride. The resistive substance is commonly a ceramic and metal alloy. To alter the film's resistance value, a laser or abrasive is used to carve a spiral pattern into it, much as one would do with carbon film. Although metal film resistors are better at handling higher frequencies, they often have worse temperature stability than their wire wound counterparts.

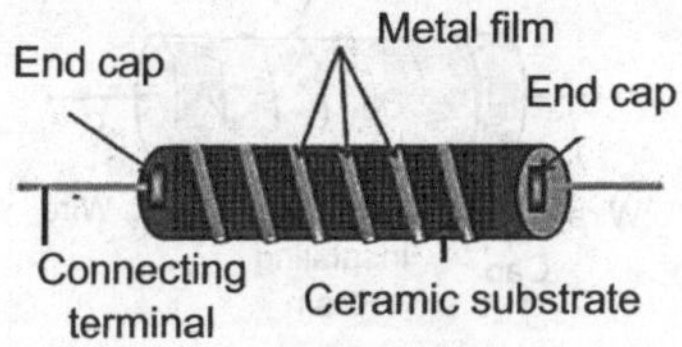

Fig. 2.5: Metal Oxide Resistors

5. **Thick Film and thin Film Resistors**

   Similar to thin-film resistors, thick-film resistors are built using a film layer of resistive material placed between two metal electrodes. Since this is the case, we refer to them as thick film resistors. These resistors are typically constructed using a ceramic rod that has a dense grid and a resistive substance. High-quality glass or ceramic may be used to create an insulating tube/tube, which can then have a thin layer of conducting material layered on top. These resistors are also divided into two categories, carbon film and metal film.

6. **Carbon Film Resistors**

   A carbon layer limits the current flow in this fixed resistor type to a predetermined value. Temperature and high-voltage applications are typical places to find these resistors in use. These resistors have a nominal working temperature range from 15kV up to 350°C. Radar, high-voltage power supply, lasers, and x-rays are prime examples.

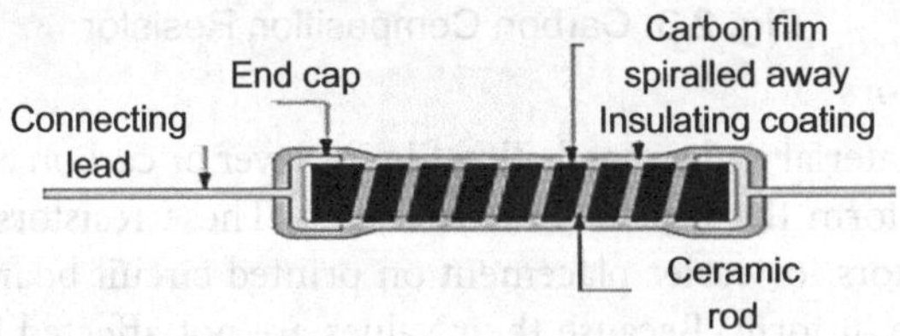

Fig. 2.6: Constructional veiw of Metal Oxide Resistors

7. **Fusible Resistors**

   These resistors are comparable to those that are wrapped with wire. A circuit's resistor may open or break if the power input exceeds a certain amount. That's why we referred to it as a "fusible resistor," then. These restores serve a dual purpose, as a current limiter and a fuse. Fusible resistors are a common component in a wide variety of audio/video equipment and electrical circuits. The resistance value of these resistors is typically less than 10 Ohms.

8. **Metal Film Resistors**

   A metal film resistor is a fixed type resistor that consists of a resistive element that resembles a thin metal sheet on top of a non-conducting substance. These resistors act like a resistance wire to provide a high tolerance.Sputtering or vacuum evaporation are common methods for forming the resistor film layer over the glass or ceramic base. The lead-type design of this resistor makes servicing and manual adjustments a snap. Metal film resistors find widespread usage in electronic and mechanical devices that operate at high frequencies, such as those used in communications, instruments, and home appliances. These resistors have a wide range of applications because to their low weight, low cost, simple design, and great accuracy. Under strict accuracy requirements, these resistors play a crucial role in the electronics and aerospace sectors.

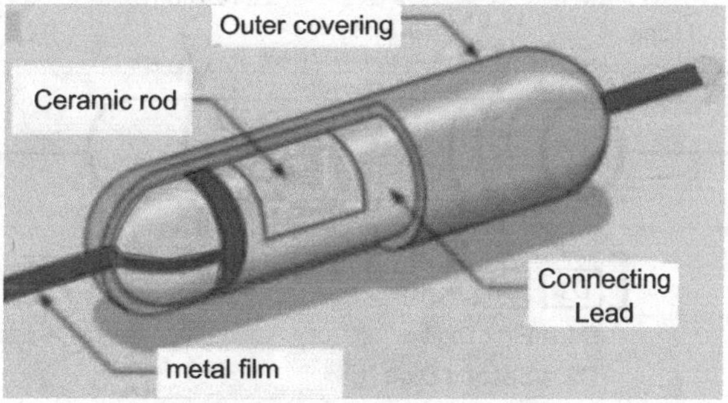

Fig. 2.7: Metal film Resistors

Resistors Colour coding:

Fixed Resistance is displayed in the form of colour coding for carbon resistors while it is written on the wire -wound resistors .The colour coding of the resistors is done in the form of coloured rings. Each colour is a code for a particular number, while the location of this coloured ring decides the placement or multiplier of the number. The colour coding is showing in given table .The method of determining the resistance value from colour coding is presented in table 2.1. The first three coloured rings determine the resistance value, while the fourth ring provides the resistor tolerance.

COLOUR	DIGIT
Black	0
Brown	1
Red	2
Orange	3
Yellow	4
Green	5
Blue	6
Violet	7
Gray	8
White	9

COLOUR	TOLERANCE
Gold	± 5%
Silver	± 10%
No colour	± 20%

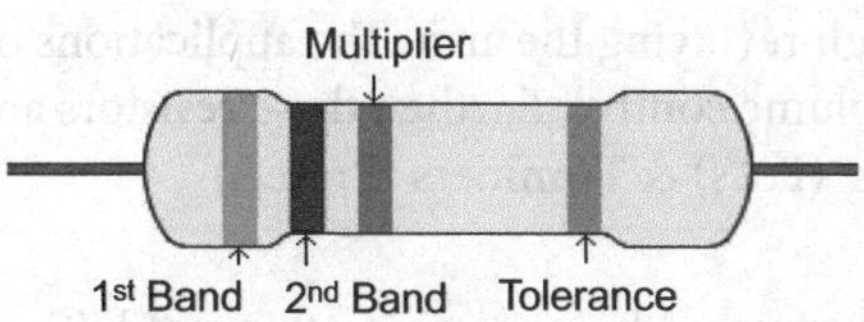

Table 2.1: Colour code and Tolerance

The multiplier is the third band. Significant numbers are determined by multiplying this factor by the results obtained from the preceding bands. The fourth band, which isn't often seen, represents tolerance when it does emerge. The resistor value is proportional to this %. The Resistor Color Code for Five Band is used to produce precise and high-quality resistors with tolerances of 1 percent, 2 percent, or less.

Besides the change in the number of digit bands, the rules are the same as they were under the prior system. Value will be indicated by the first three bands, multiplier by the fourth, and tolerance by the fifth.A five-band resistor has the colours Black, Brown, Black, Red, and Brown; hence, its estimated values are 0/1/0/102 = 10 * 100 = 1K ohm/1K, with a tolerance of 1% for Brown

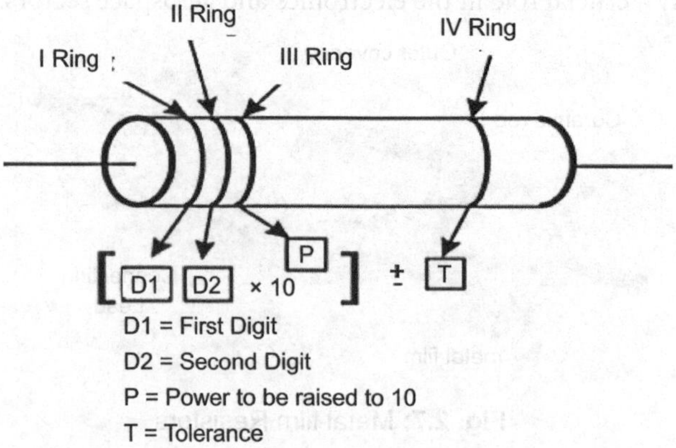

Fig. 2.8: Four Band Resistor color coding

For this reason, a high-precision resistor has a value between 900Ω and 1.01 KΩ, and its color coding will likely consist of six hexadecimal numbers. The resistance value may shift by 1000 ppm = 0.1% for every 10 C change in temperature, which is why the sixth band is often colored with brown. It is printed on the bodies of wire wound resistors what their resistance is; for instance, an orange-red-brown-brown-green-red six-band resistor has a resistance of 3.21 kΩ with a tolerance of ±1% and a temperature coefficient of 50 ppm/°C.

2K2 means 2.2kΩ

2E2W means 2Ω /2 watts

6K2 means 6.2kΩ.

- **Variable Resistors**

  A variable resistor is a resistor whose resistance value may be changed. These resistors include a rotating shaft as well as a wiping contact. When the resistance of this resistor has to be adjusted, the sliding contact may be moved along a straight line on top of the resistive element, making it a linear sliding-type resistor. These resistors include a sliding arm, which is attached to the shaft as well as the resistance value that can be changed through revolving the arm. The applications of these resistors include the radio receiver to control the volume control. Further, these resistors are classified into three types like Rheostats, Potentiometer 's (Pot's) & Trimmers (Presets)

- **Rheostat**

  In order to control current, a rheostat, which is a kind of variable resistor, is often used. They can seamlessly alter the resistance in a variety of circuit configurations. Rheostats are similar to potentiometer's in function, however their design only involves two connections rather than three terminals. The first link is linked to the resistive element's terminal end, while the linked page is linked to the wiper, the sliding contact. As compared to potentiometer, rheostats must carry current. Thus they are typically designed like wire wound type resistors where the resistive wire can be wounded approximately an insulating ceramic core & the wiper slides on the winding's.

These are frequently used like power control devices for controlling light intensity, motors speed, ovens, etc. Rheostats can be changed with switching electronics within power control applications

- **Potentiometer**

  An alternative name of a potentiometer is a pot meter/pot. It is a three-terminal resistor where the resistance can be changed manually for controlling the current flow. The construction of a potentiometer can be done using different materials like cermet, wire wound, carbon composition and metal film/ conductive plastic. This functions as a variable voltage divider. The resistive element may have two terminals linked to either end, and a third terminal that is connected to a wiper, a sliding contact that glides across the resistive element.The o/p voltage of this resistor can be determined through the wiper position and this resistor works as a variable voltage divider. These resistors are used in audio volume control & many other applications.

- **Trimmer**

  Trimmer has an extra screw along with variable resistors or potentiometer for better operation & efficiency. Here, this screw is used to change the resistance value by changing its position to revolve through a small screwdriver. These resistors are built with different materials like carbon film, carbon composition, cermet & wire materials which are accessible from the 50 Ohms to 5 mega ohms range. The power rating of these types of resistors ranges from 1/3 to ¾ Watts.Trim pots dont require more adjustments in their complete lifetime. A trim pot usually calibrated in initial after manufacturing .Trim pots are basically used in A/V components and may require adjustment when the equipment is being serviced.

2.2.2 Non-Linear Resistors

Nonlinear resistors are those whose voltage and current characteristics may be altered linearly. Light, temperature, and other environmental factors will all affect the voltage and current levels. Varistors, thermistors, Light Dependent Resistors (LDR), and surface-mount resistors (SMDs) are just a few examples of the many subsets into which non-linear resistors fit.

- **Voltage Dependent Resistor (VDR)**

  An other name for VDR. is varistor. The resistance of this resistor is adjustable & relies on the applied voltage. Once the applied voltage rises or greatly increases then the resistance will be substantially decreased or normally decreased. Therefore, this behaviour is crucial for protecting circuits from voltage surges since failure to do so might result in lightning strikes & electrostatic discharges. The metal oxide varistor (MOV) is the prototypical example of this kind of resistor. Varistors are used for control or compensation the elements in circuits it may provide optimal operating conditions or to protect against excessive transient voltages

- **Thermistor**

  A thermistor is a resistor whose resistance varies mostly with temperature. Combines thermal and resistive elements into a single device.A thermistor may be made from metallic oxides by pressing them into a bead or a disk and then encasing them in a tough material like glass or epoxy to protect them from the environment. Thermistor types include PTC (positive temperature coefficient) and NTC (negative temperature coefficient) .Resistance will decrease in the NTC type as temperature increases. In a similar vein, the resistance increases as the temperature decreases.As a result, this thermistor type sees the most widespread use. PTC devices function differently from NTC models because their resistance also rises with rising temperature. Similarly, the temperature falls as resistance reduces. Accordingly, this kind of resistor is used in the function of a fuse.

- **Light Dependent Resistor(LDR)**

  According to working of LDR once light falls on this resistor, the resistance will be changed based on the light intensity. This type of resistor can be made through cadmium sulfide which includes fewer electrons once it is not light up.Once a light signal drops on LDR, then electrons will get ejected, thus its conductivity will be increased. Therefore, it gives less resistance once the light signal drops on LDR and provides high resistance within the dark. LDR are used as light sensors and alarm clocks, burglar alarm circuits,street lights, light intensity meters,
- **Surface Mounted Device(SMD) Resistor**

  SMD resistor is one kind of electronic component where SMD stands for (SMT) Surface Mount Technology allows this resistor to be mounted directly into a PCB circuit. These resistors often come in smaller sizes than the conventional variety, allowing for more efficient use of PCB real estate. Surface mount devices was invented mainly for reducing the size of the components & also the time it takes to design a circuit.

2.2.3 Other Types of Resistors

Below is a discussion of several other resistor kinds, including pull-up, shunt, load, magneto, and load.

- **Magneto Resistor**

  Magneto resistors have resistance that varies with the intensity of the magnetic field. This resistor's primary use is in gauging the presence, direction, and intensity of magnetic fields. Therefore, these resistors are essential for detecting and measuring magnetic fields.Magnetometers, which measure magnetic fields, make use of magnetic dependent resistors (MDRs). This resistor's resistance varies with the direction and intensity of the magnetic field.
- **Load Resistor**

  In the context of developing and testing circuits, the load resistor is one sort of component that serves as the ideal output. These resistors are useful for transferring as much power as possible, ensuring a stable output via impedance matching, and limiting current flow to a minimum. The usage of these resistors is essential to the functioning of a switching power supply.
- **Shunt Resistor**

  Low-resistance resistors are known as shunt resistors. To create this resistor, a material with a lower TCR (temperature coefficient of resistance) is possible. This resistor may be connected in series with the load whose current is being measured, or in parallel with the ammeter. In order to measure the current in a circuit, shunt resistors are often used.
- **Pull Up Resistor**

  Pull-up resistors are used to drive a wire to a high logical level in the absence of an input signal. Used in logic circuits, these resistors maintain a consistent logic level at a pin no matter the environment.

2.2.4 Testing of Resistor

In reality, the resistance of resistors typically ranges from 1 ohm to 10 meg-ohm. An analogue multimeter or an ohmmeter may be used to measure resistance, and the results can be compared to the one indicated by colour coding (Fig. 2.4). The resistor is accurate if its measured value is the same as, or close to, the value computed using the color code.

Measurement of Resistance:

The steps involved in measurement of the resistance are as follows.

1. Switch 'ON' the multi-meter and point selector switch in 'Ω' section.
2. Now connecting lead should be connect to 'V/Ω' and 'COM' sockets
3. Selector switch should be at lowest range in 'Ω'
4. Connect the tips of the leads to both the leads of the resistor.
5. If the resistance of the resistor is in the selected range ,then,the resistance value will be displayed,otherwise it will shows "over-range".
6. At over-range display position ,the selector should be switch at next higher value and repeat till value is displayed.
7. For an open resistor "over-range" will be displayed in even the higher selected range.
8. The range can be also be directly selected depending on the colour code values

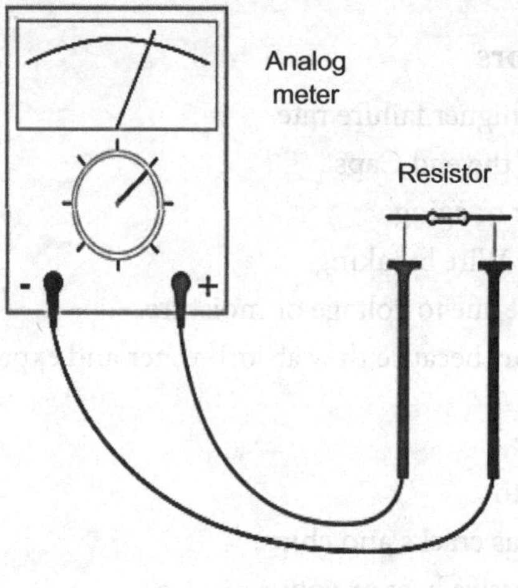

Fig. 2.9: Testing method for fixed resistors

The resistance scale of digital multi-meters is linear, making them more convenient to use. They use a source of steady current to generate a voltage across the unknown resistance; this voltage is then measured, and its value may be deduced from the known current through the resistor. The constant current used is of a value that varies with the working range. Digital multimeters can detect very low resistance values, such as the contact resistance of plug and socket connections, according to the four-terminal (Kelvin) principle. To facilitate the taking of these measurements, a larger current is used. This causes a noticeable voltage decrease.

Precautions:

1. At the time of measurement do not touch the resistor leads with hand because body resistance can effect the original value
2. The multi-meter leads should be touched properly to the resistor leads to minimize the contact resistance.
3. While making measurement with the help of analog multi-meter ,zero ohm adjustment should be made before starting the measurement process and also when a new resistance range is selected

4. If oxide layer is there on the lead of the resistor scratch it off with some start to or with the help of the science paper oxide layer increases the resistance measured with multi-meter

2.2.5 Failures in Fixed Resistors

An very low failure rate is typical with resistors because of their dependability. Resistors, however, are not immune to failures. Due to chemical or other changes brought on by heat, applied voltage, humidity, mechanical stress, and vibrations, ageing causes progressive deterioration in resistors.

The maximum amount of power a resistor can dissipate when added to a circuit is inversely proportional to the temperature of the surrounding environment, hence it is crucial to consider the temperature of the circuit before adding the resistor. Reduced dissipation naturally results in greater stability and reduced failure rate figures. A resistor's failure rate is generally proportional to its kind, manufacturing technique, operating and environmental conditions, and value. Possible failures of fixed resistors and their explanations include the following.

Carbon Composition Resistors

These type of resistors are having higher failure rate

1. Due to heat or dislodging of the end Caps .
2. Due to environmental factor or aging
3. Due to excessive flexing and Wire breaking
4. Carbon or binder movement due to voltage or moisture.
5. Carbon fragments break apart because they absorb water and expand.

Know more about color coding of resistors

Metal Film Resistors

1. Metal film resistors fail due to
2. Film production flaws such as cracks and chips;
3. Film breakdown due to excessive heat or voltage
4. Due to the thin resistance spiral, higher value resistors are more prone to open circuit failure.
5. High noise
6. Mechanical stress from poor circuit construction can lead to poor contact at the end connectors.

Wire Wound Resistors

Wire wound resistors fail due to

1. Wire breaks, especially small diameter wire.
2. Electrolytic action from absorbed moisture causes wire corrosion;
3. A wire's slow crystallization (due to impurities) can create breaks and breaks in continuity.
4. The welded end has a disconnect.

The percentage change in resistance value over time is a measure of a resistor's stability. That is dependent on the rate of heat production and the temperature of the surrounding environment. The hot spot temperature of a resistor is equal to the combination of the ambient temperature and the temperature increase induced by power dissipation. The highest temperature, or hot spot temperature, is located smack in the centre of the resistor body thanks to its homogeneous design.

2.2.6 Variable Resistors

A resistor with a variable amount of electrical resistance can have its resistance value changed on demand. As a result of Ohm's Law, voltage and current may be regulated by adjusting the resistance of a circuit, and variable resistors are commonly employed for this purpose. Sliding a wiper contact along a resistance track changes the electrical resistance. In some cases, the resistance is set to a specific value when the circuit is built using an adjustment screw, while in other cases, the resistance is set to a default value and may be changed as needed using a control knob. It is the location of the slider contact along the resistance track that determines the effective resistance value of the variable resistor. A resistance track and a wiper contact make up the bulk of its construction. When the movable part is adjusted, the wiper contact slides along the resistance track. Carbon track, cermet (ceramic and metal combination), and wire wrapped track are the three primary forms of resistance track employed in this resistor. High-resistance applications benefit from carbon track or cermet track, whereas low-resistance variable resistors benefit from wire wrapped track. In most circumstances, the resistance tracks will be circular in form, while in others, a straight track will be used

2.2.7 Variable Resistor Connection

When just one terminal of the resistance track and the wiper terminal are connected to the circuit, the variable resistor functions as a rheostat; the electrical resistance between the connected terminal and the wiper terminal varies with the position of the wiper on the resistance track. When both terminals of the resistance track are connected to the input circuit, the device functions as a potentiometer; the resistance between the connected terminal and the wiper terminal stays constant. An adjustable resistance may be necessary in an electronic circuit, however this modification may be needed just once or often. Preset resistors are wired into the circuit to achieve this result. One kind of variable resistor is the preset resistor, whose electrical resistance value may be adjusted through a set screw.

2.2.8 Variable Resistors as Potentiometer

Variable resistors basically consist of a track of some form of resistance material to which a movable wiper makes contact. There are three distinct types of variable resistors that are differentiated by the their resistive material (Fig.2.10).

- Carbon: Carbon potentiometer are formed of either molded carbon composition giving a solid track or a coating of carbon with insulating filler onto a substrate.
- The Cermet potentiometer has a thick film resistance layer on a ceramic base.
- A wire-wound potentiometer is made by wrapping Nichrome or another resistance wire around a suitable insulating former.

Types of Potentiometer

Potentiometer may be classed into the following categories based upon the amount of resistors and the control arrangement utilized:

Single Potentiometer: With a single resistor, a potentiometer can be used for control.;

Tandem Potentiometer: Two identical resistor units controlled by one spindle;

Twin Potentiometer: Consisting of two resistor modules with separate control spindles

Multi-turn Potentiometer: Resistance may be adjusted by a potentiometer, which can have anything from one to forty spindle rotations.

Pot pack: One-turn or multi-turn rectangular potentiometer.

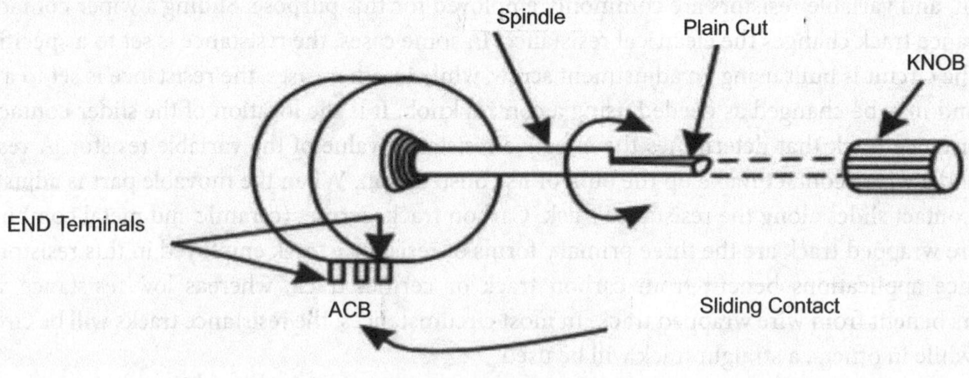

Fig. 2.10: Variable resistors

Typically, a potentiometer will be used to regulate the gain of an amplifier, set the time constant of an RC timer, set the bias of a transistor, or transport current or voltage in a control circuit. This is why the components come in PCB-mounting-application-friendly packing.

Rheostats and potentiometer s both use variable resistors. The distinction between the two uses is seen in Figure 2.7. A voltage divider has a resistor element that is linked to a voltage reference source and a sliding arm that serves as the pick-off point and can be adjusted to get the desired voltage.

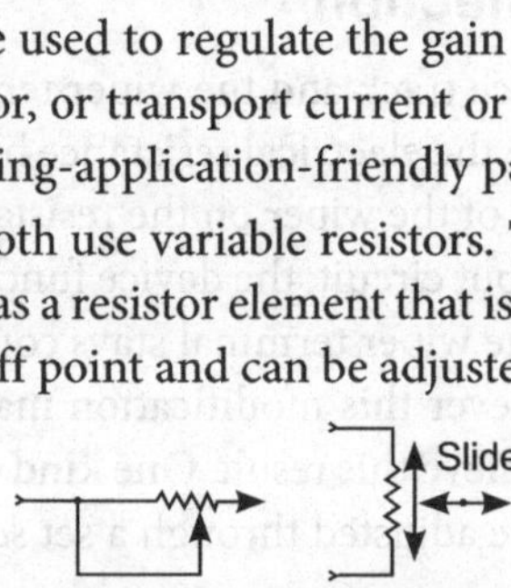

Fig. 2.11: Difference between a potentiometer and rheostat operation

The resistor element in a variable resistor design has one end linked to the circuit and the other end attached to the sliding arm. The slider may also be linked to an external circuit, with the full resistance being in series. A potentiometer-like function is still feasible in this configuration.

The following are some of the many design variables for variable resistors: Every part of the length of the pot presents the same amount of resistance, making it linear.

Log: The logarithmic rule is reflected in the pot's resistance, which changes with time. When the wiper is rotated, the resistance in these pots rises very gradually from zero to about halfway. At that point, the resistance will grow at a significantly faster rate relative to the first half of the pot-rotor revolution when the wiper shaft is cranked farther.Sine-cosine With a potentiometer, the wiper's resistance varies along the track according to the sine-cosine law, as the name suggests. The complete length of the operating track, which covers all 360 degrees of rotation, is split into four equal 90-degree segments.

Working Principle of a Potentiometer

It basically works on the principle of varying the resistance of a fixed resistor by moving the sliding contact of the wiper. By displacing the position of the wiper, the length of the resistive constantan or manganin wire is altered. This change in the length of the wire is proportional to the change in the resistance offered to the circuit. Also, the resistance is inversely proportional to the cross-sectional area of the wire; however, it is difficult to vary the cross-section of a wire. Hence, the cross-sectional area of the wire and the current flowing in the circuit is maintained constant and the length of the resistive wire is made variable. In a potentiometer, input is applied across the fixed terminals, whereas the output is noted between the movable and a fixed terminal.

2.2.9 Failures in Potentiometer

Compared to fixed resistors, potentiometers have a greater failure rate. This is due to the fact that both the wiper and the resistance track are mechanically moving components that need a consistent and solid electrical contact to function properly. There are often two categories of problems that arise during effective application:

Complete Failure: The open circuit between the wiper and track, or the track and the end connectors, is often symptomatic of a catastrophic failure.

This can be caused by:

1. Metal components deteriorate due to humidity.
2. plastic item (track moulding) swelling and distorting because of excessive humidity or temperature.
3. A potentiometer's wiper contact resistance may increase, leading to an increase in electrical noise, or the contact may become intermittent if it fails in part. The problem arises when dust, grit, or oil becomes wedged between the wiper and the track. Noise in an audio circuit, strange behavior of the regulated parameter, etc. are all signs that a potentiometer's contacts are deteriorating.

2.2.10 Testing of Potentiometer

A potentiometer's functionality may be tested using an ohmmeter in the same manner as fixed resistors are. The terminal with the changeable contact and the two fixed terminals should be tested separately. Figure 2.8 It is possible to monitor the resistance while the pot is turned by turning the shaft. If the resistance reading persists at zero or infinity, you may throw out the pot and get a new ones. The resistance value, tolerance, power rating, resolution, form, size, and location of the adjustment screw of the new potentiometer you choose to replace the old one must all be correct.

Know more about potentiometer

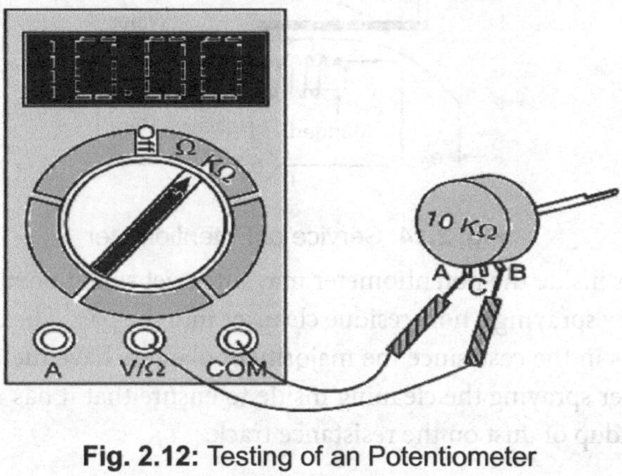

Fig. 2.12: Testing of an Potentiometer

Procedure of testing a potentiometer using digital multi-meter

1. Determine the Rating of the Potentiometer

   This rating is reflective of the potentiometer's set ohm value between 1K to 10K.

2. Set up the Digital Multi-meter

   Connect the red jack to the port and the blackjack to the COM port to switch the multimeter to resistance mode. Follow that by setting the dial to the Ω symbol.

3. **Inspect the Potentiometer Closely**

   Now, let's take the potentiometer and give it a good once over. There must be a total of three terminals: two end terminals and a wiper terminal. Before conducting a resistance test, it is necessary to correctly identify each terminal,

4. **Check the Reading**

   Then, check the reading. and this reading should not change while you rotate the controller.

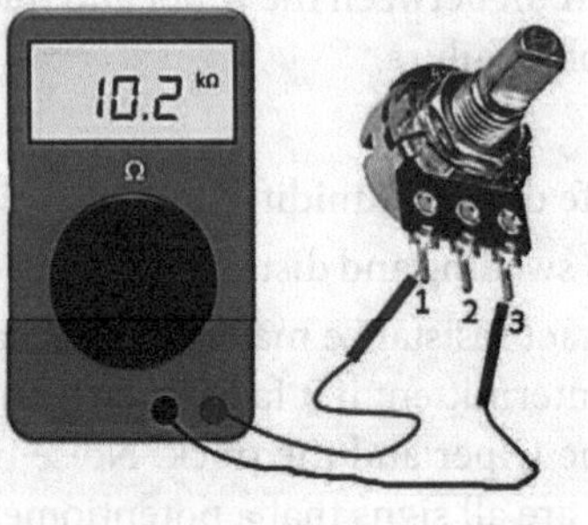

Fig. 2.13: Potentiometer reading checking

5. Test the Wiper

   The next step is to move one of the alligator clips from the end terminals to the wiper terminal. We can test this by spinning the controller and seeing whether the reading on the multimeter changes when we do so. and then use the controller to test the lowest and maximum resistance.

2.2.11 Servicing the Potentiometer

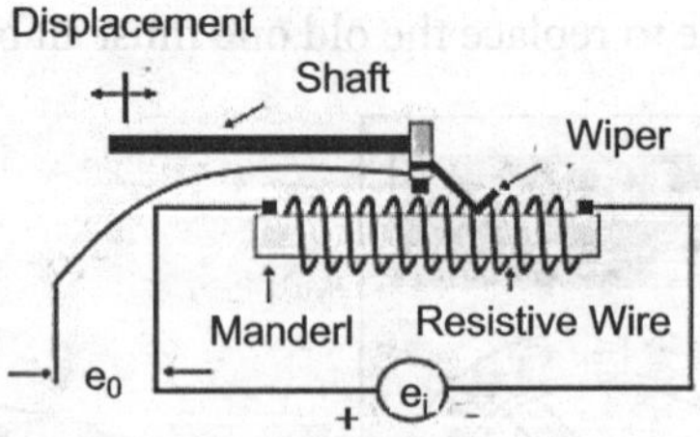

Fig. 2.14: Service of Potentiometer

The wiping contacts inside the potentiometer may often get soiled or rusted over time. Most of the time, this is easily fixed by spraying a non-residue cleanser into the pan. The cleaner is sprayed down the shaft or into other cracks in the case since the majority of planters have metal casings.You need to spin the shaft many times after spraying the cleaning inside to ensure that it has reached every crevice. This would eliminate the buildup of dust on the resistance track.

2.3 LIGHT DEPENDANT RESISTORS (L D R'S)

Cadmium sulfide is used to create LDRs. When stored in total darkness, there are extremely few free electrons for them to interact with, resulting in a very high resistance. Light may free electrons, making a substance more electrically conductive. When the power is turned off, the caught electrons are released back into the material, making it an insulator rather than a conductor. An LDR typically has a dark resistance between 1 M ohm and 10 M ohms. Between 75 and 300 ohms is its mild resistance. The amount of time it takes for the LDR to reach this new condition is referred to as the recovery period. Commonly, it is 200 Kohm/sec for the recovery rate.

Light-dependent resistor (LDR) testing is straightforward, requiring just the measurement of resistance in the dark and in the presence of light, and may be performed using a voltmeter or ohmmeter.

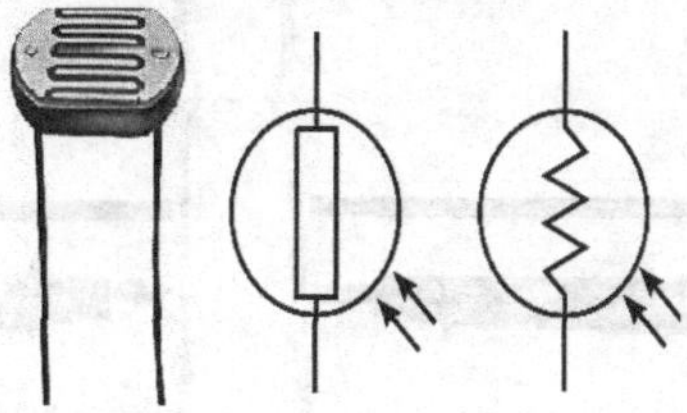

Fig. 2.15: Symbol of Light Dependent Resistor

2.4 THERMISTORS

Thermistors are special kind of resistors having a very large resistance-to-temperature coefficient. Most commonly used thermositors have a negative temperature co-efficient, meaning their resistance value decreases as the temperature rises. These substances are oxides of metals like nickel, cobalt, and manganese. Thermistor come in a wide variety and may be used for many different purposes.Disks, beads, and rods of various diameters and thicknesses are included in the assortment. The following diagram depicts the thermistor symbol used in both the United States and abroad.

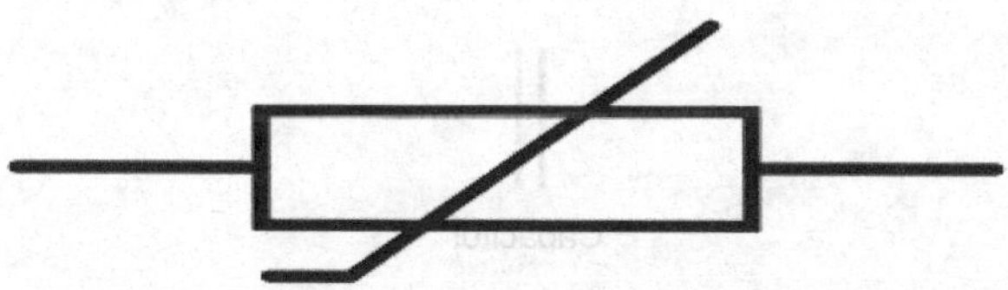

Fig. 2.16: Symbol of Thermistor

In a thermistor with a negative temperature coefficient (NTC), the resistance decreases temperature increases, while a thermistor with a positive temperature coefficient (PTC) experiences the reverse. To the contrary, resistance rises as temperature drops. This type of thermistor is used the most.

The operation of a PTC thermistor differs slightly from that of a standard one. The resistance goes up as the temperature goes up, and it goes down as the temperature goes down. This thermistor is often used as a fuse. Within a very narrow temperature range, typically approximately 50°C around the target temperature, the thermistor is able to attain excellent accuracy. The base's resistance determines this band's usable temperature. Positors are thermositors having a positive thermo-resistive co-efficient. Having been crafted from barium titanate ceramic, they are distinguished by a huge shift in resistance over a very narrow temperature range. Various appliances, including electric motors, washing machines, alarm systems, etc., employ thermistors for current limiters, temperature sensors, and overheating prevention. They may also be used as compensation resistors, time delay devices, and thermostats. Depending on the application, thermistor beads must be shielded by being either ceilinged into the tip of a glass tube or enclosed inside a stainless steel sheet. A digital multimeter (DMM) or voltage ohm meter (VOM) may be used to quickly and easily test a thermistor by measuring its resistance both at room temperature and when heated by a soldering iron tip, a blow dryer, or a heat gun.

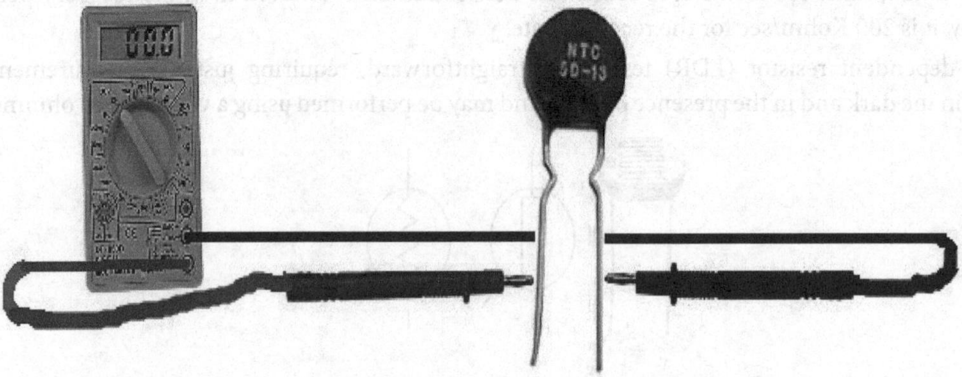

Fig. 2.17: Thermistor Testing

The resistance should gradually grow or decrease whether it is PTC or NTC. The gadget is defective if the resistance displays erratic behavior, such as going to infinity or zero. However, you'll need specification and temperature measurement devices to verify the thermistor's functionality.

2.5 PASSIVE COMPONENT: CAPACITOR

A capacitor can store energy as an electrical charge, thus the term. Essentially, it is made up of two metal plates that are isolated from one another by a dielectric

Know more about capacitor

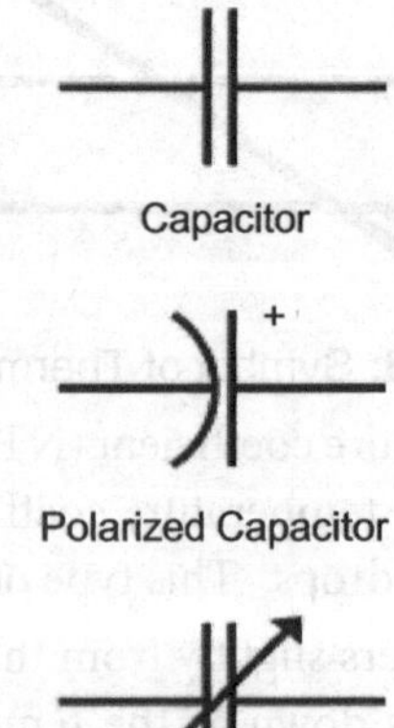

Fig: 2.18: Basic Capacitor Symbol

This insulating substance may be made of wax paper, mica, ceramic, plastic, or a liquid gel similar to that used in electrolytic capacitors. The insulating layer between the capacitor plates is called the dielectric. The kind of capacitor is determined by the dielectric. Capacitors come in a variety of materials, including mica, paper, polyester, ceramic, and electrolytic. Capacitor voltage ratings are all over the map. The capacitance of a ceramic capacitor is often in the picofarad range, if that. A capacitor made of electrolyte may store several microfarads of energy. The breakdown voltage of the dielectric used between the plates determines the maximum operating voltage. When a dielectric in a capacitor reaches a certain voltage, it stops being insulating and becomes conducting, this value is called the breakdown voltage. The capacitor's plates get short-circuited as a result of this.

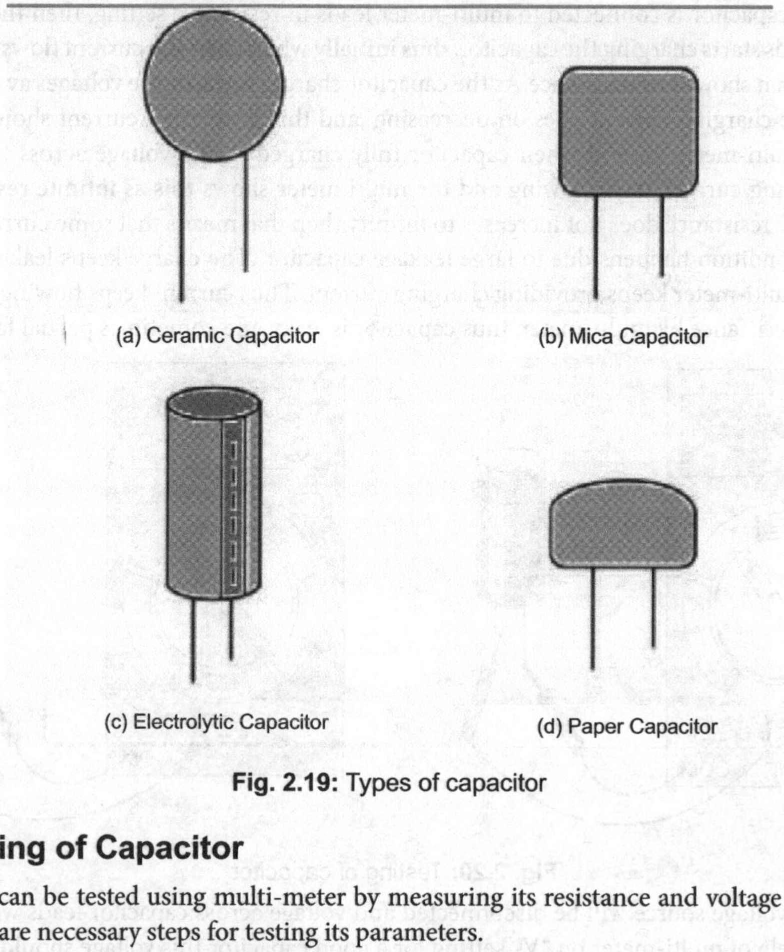

Fig. 2.19: Types of capacitor

2.5.1 Testing of Capacitor

- Capacitor can be tested using multi-meter by measuring its resistance and voltage after charging. Following are necessary steps for testing its parameters.
- Firstly set multi-meter to its maximum range using selector switch.
- Then connect probe tips to any leads of capacitor but that connection not applicable in electrolytic capacitor because in this particular capacitor black lead should be connected to negative lead of capacitor and red lead to other
- After connection set up the multi-meter should show increasing resistance,which increases ultimately to attain over range for a good capacitor.

- In analog multi-meter pointer firstly deflecting to right,and then will return back to left that calibration indicating increasing resistance. For a good capacitor resistance should be increases to infinite .
- If multi-meter not having increasing value of resistance shows low or approximately zero value that indicating that capacitor is partially or completely short respectively
- If capacitor will be open than multi-meter indicates open circuit.
- If resistance of capacitor increases but not reaching over range or maximum infinite value than this situation indicates leaky capacitor

When the capacitor is connected to multi-meter leads in resistance setting, then the voltage in the multi-meter leads starts charging the capacitor. thus initially when charging current flows through multi-meter leads,than it shows low resistance.As the capacitor charges towards the voltages available in multi-meter leads the charging current goes on decreasing ,and this decreasing current shows as increasing resistance by multi-meter. In end when capacitor fully charged to the voltage across the multi meter leads, the charging current stops flowing and the multi meter shows this as infinite resistance or over range. When the resistance does not increases to infinity then that means that some current still flowing in leads. This condition happens due to large leakage capacitor .The charge keeps leaking through the capacitor and multi-meter keeps providing charging current. Thus current keeps flowing ,which in term showing finite resistance by multi-meter. thus capacitor is leaky and sometimes partial failure

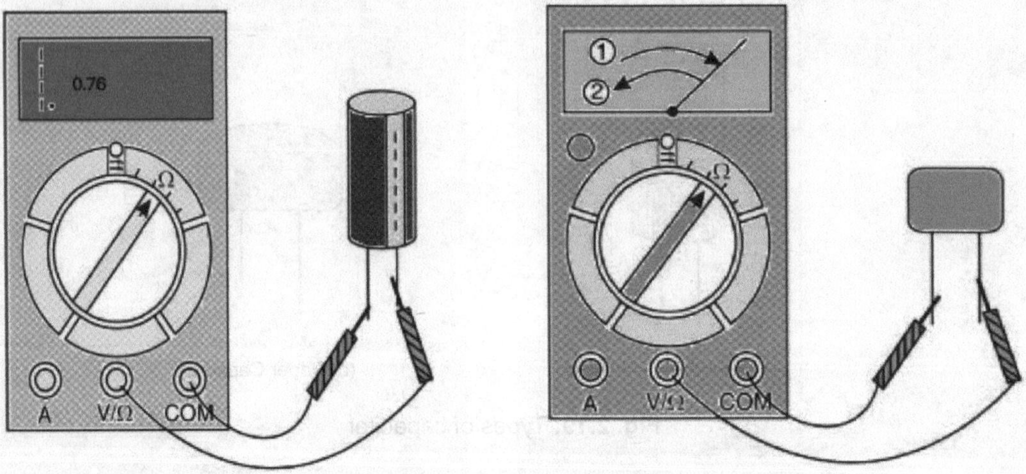

Fig. 2.20: Testing of capacitor

- Than DC voltage source will be disconnected and voltage across capacitor leads will be measured with the help of multi-meter on "V-"setting,for a good capacitor this voltage should be equal to the DC source across which the capacitor was connected .

If the multi-meter measure voltage less than the DC source voltage which the capacitor was connected,then it means that the capacitor is leaky

$$Q = CV$$

$$Q \times V$$

$$V \times Q$$

The charge 'Q 'and Voltage 'V' of capacitor are related as

The above relation shows that the voltage across a capacitor directly promotional to charge,and vice versa.Therefore,we can test the capacity or charge retaining capability of capacitor by measuring voltage across a charged capacitor.so we can say initial applied voltage is in low value means leaked charge

2.5.2 Charged-Voltage Testing

Capacitor can also be tested by measuring the charged voltage across the capacitor.For this method,following steps are followed.

- Place the capacitor's leads across a DC voltage source whose output is lower than the capacitor's maximum rated voltage.

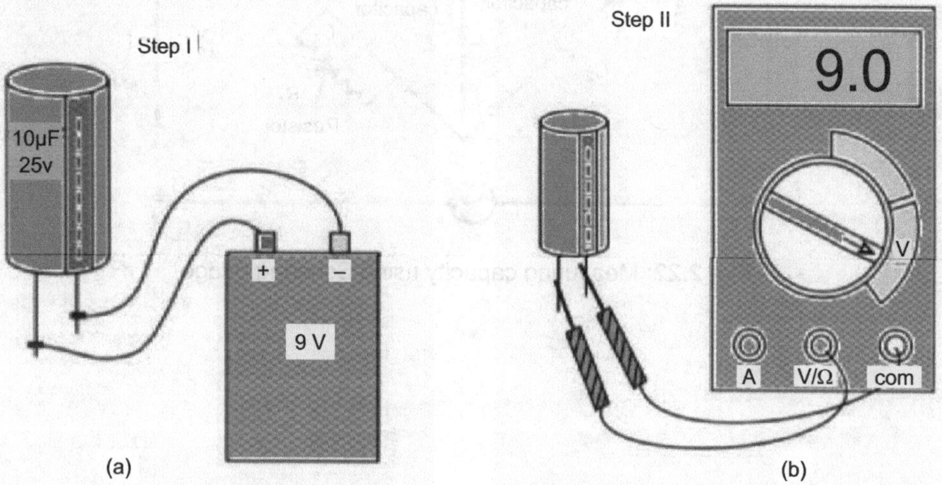

Fig. 2.21: Charged voltage testing of capacitor

- To ensure the capacitor charges to the full DC voltage of the source, we will now let the capacitor to stay in contact for a few seconds.

2.5.3 Failure in a Capacitor

The possible faults in the capacitor can be "open",short or "leaky"behaviour as explained below

- The capacitor can get open circuit due to not proper leads connections from inner plates. This may be due to mechanical stress or disconnection of the welding due to excess heat generation.
- Due to high applied voltage capacitor can be damaged and causing burning due to excess heat generation
- Due to dielectic breakdown ,high voltage,and enviornment factors ,inter connection of the metal plates,or due to breakage in "in-between"dielectric material or misalignment at the ends capacitor can get short
- The capacitor can get leaky due to degradation of the dielectric properties of the dielectric with aging, over-voltage application,environmental factor like humidity, or substandard quality of the material.

2.5.4 Measurement of capacitance

The capacitance of a capacitor can be measured with the help of many RLC meter, Schering bridge etc. The capacitance can be measure by connecting the capacitor in the bridge and balancing the bridge.

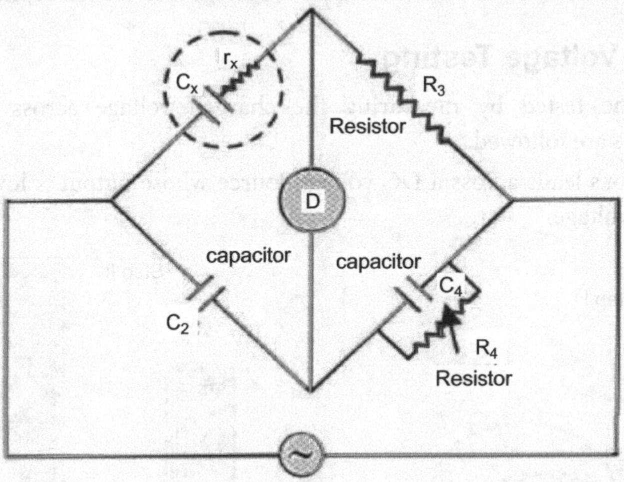

Fig. 2.22: Measuring capacity using Schering bridge

$$C_x = \left(\frac{R_4}{R_3}\right) C_z$$

Many new models of digital multi meters have the facility for the measurement of capacitor.Such multi-meters have small opening slots so as to enable insertion of the capacitor leads. The front panel of the multi-meter has a section of capacitance measurement also for the selector switch.

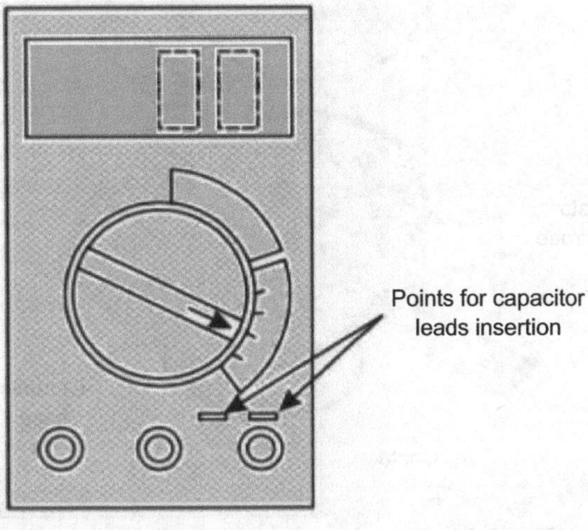

Fig. 2.23: Capacity measurement using multi-meter

The procedure is to insert the leads of the capacitor, capacity value of the capacitor will be displayed. It inadvisable to start from the highest range ,and then proceed towards lower range till the capacity value is displayed in sufficient significant digits on the display.The disadvantage of measurement of capacitance with the method is that the leakage-present in capacitor can not be found out.

2.5.5 Types of Variable Capacitor

The following two configurations of variable capacitors are often offered:

- Button type has a variable rotor
- Tubular type has an adjustable core

It should be noted that changes made to variable capacitors using a metal screwdriver are affected when the screwdriver is taken from the rotating screw. This is due to the fact that turning this screw with a metal screwdriver alters the stator's or, more often, the rotor's, effective area of the metal plated surface. Use of a nonmetallic screwdriver is advised in this situation.

Any one of the dielectric materials, such as ceramic, mica, polystyrene, or teflon, may be used to build variable capacitors. A variable capacitor essentially consists of a stator and a rotor. The stator's area is constant, but when the rotor is rotated from 0 to 180 degrees, the amount of the plate's surface exposed vary, changing the capacitance.

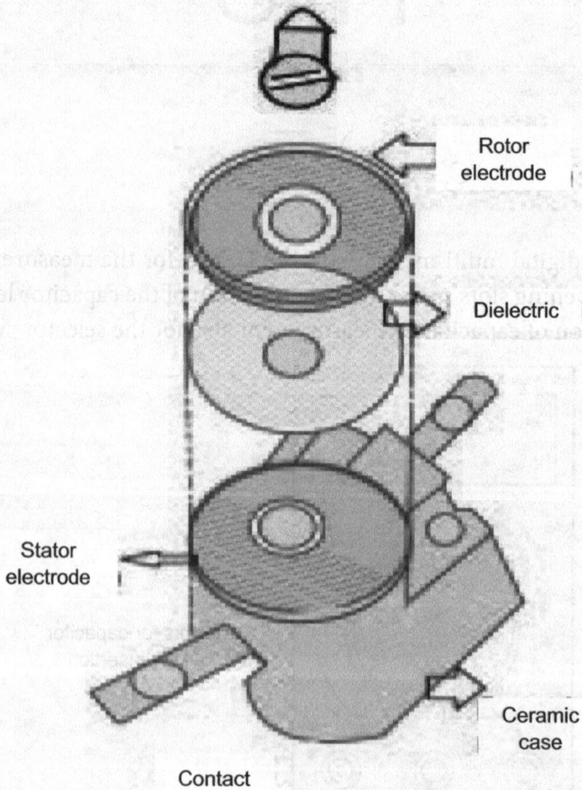

Fig. 2.24: Button type variable rotor capacitor

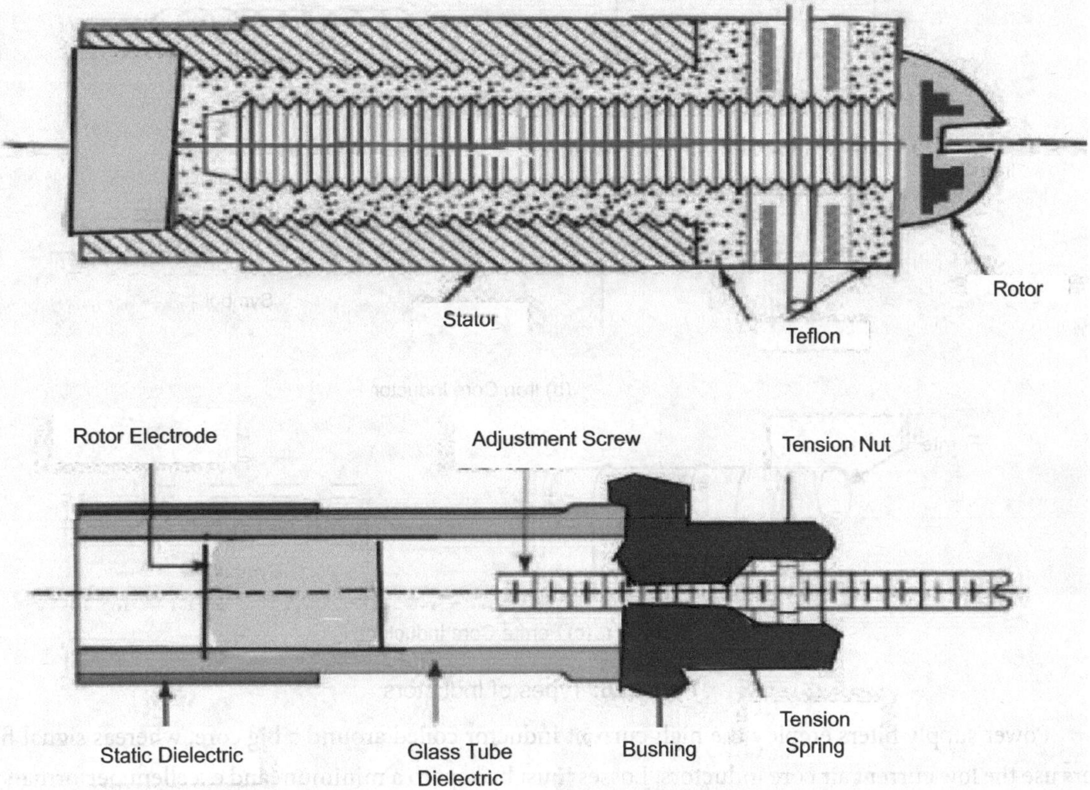

Fig. 2.25: Tubular type variable capacitors

2.6 PASSIVE COMPONENT: INDUCTORS

A device's inductance is the degree to which it opposes a modification in the current flowing through it.

Inductors function according to the 'self inductance' principle, which states that any change in the current passing through a coil of wire causes the inductor to generate a magnetic field. The DC current is non varying,so the inductor is useless in DC circuits,its only useful in AC circuits. The construction of an inductor consists of a winding or coil of usually copper or aluminium.The inductor are categorised depending on the type of core on which the coil is wound .The inductor are categorised depending on the type of core on which the coil is wound. The cores used are soft iron,ferrite or no core. Accordingly,inductor are named Iron-core inductors, Ferrite core inductors and air core inductors respectively. The iron core is usually made of thin strips or laminates in order to reduce eddy currents Inductance is measured in henries. A one-henry inductance will produce a one-volt counter emf at a current of one amp per second (electromotive force). Like a resistor, an inductor's value may be inscribed on the component's body or shown by colour bands (Fig. 2.20). The inductor primarily serves as a filter. There are two distinct characteristics of filter inductors.

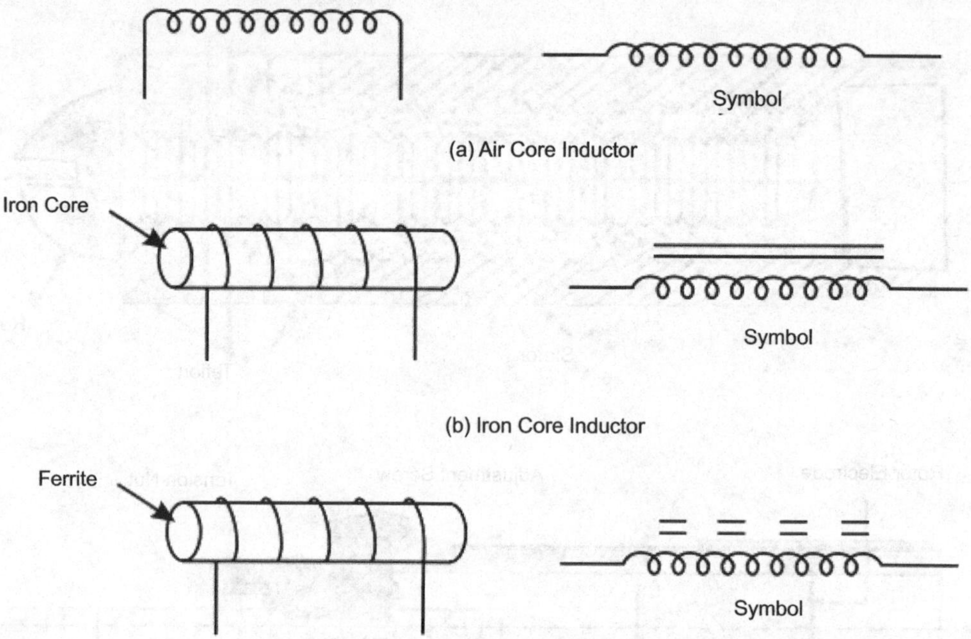

(a) Air Core Inductor

(b) Iron Core Inductor

(c) Ferrite Core Inductor

Fig. 2.26: Types of Inductors

Power supply filters employ the high current inductor coiled around a big core, whereas signal filters use the low current air core inductors. Losses must be kept to a minimum and excellent performance must be attained in high current inductors; this is why cores are being used. The big size and weight of the cores are due to their exceptional density. Huge inductors or transformers are often used in switching power supply to reduce switching noise and smooth out the voltage waveform at the load. In signal processing circuits, low current inductors are employed as filters. Since the inductive/capacitive filter's slopes are steeper than those of the resistive/capacitive filter, it is superior in certain situations. Inductors are often only seen in electrical power circuits.Inductors are two types fixed and variable types .The construction difference is the movement of core that is movable or fixed.The core is provided with screw facility ,so that it can be moved axially by rotating it.The winding of the inductor is insulated so that there is no lateral or side-by-side electrical connection of the turns. usually ,an insulating paint is applied on the wire before winding it.This paint avoids the lateral electrical connection of current flows in the inductor

Know more about Passive components

Table 2.2: Colour coding for inductor

Inductor Band Colour Codes

1st Band (Value)		2nd Band (Value)		3rd Band (Multiplier)		4th Band (Tolerance)	
		Black	0	Black	× 1 or no zeros		
Brown	1	Brown	1	Brown	× 10 or + 1 zero		
Red	2	Red	2	Red	× 100 or + 2 zeros	Red	± 2%
Orange	3	Orange	3	Orange	× 1k or + 3 zeros	Gold	± 5%
Yellow	4	Yellow	4	Yellow	× 10k or + 4 zeros		
Hreen	5	Hreen	5	Green	× 100k or + 5 zeros	Green	± 5%
Blue	6	Blue	6	Blue	× 1 m or + 6 zeros	Blue	± .25%
Violet	7	Violet	7			Violet	± 1%
Grey	8	Grey	8	Gold	× .1	Gold	± 5%
White	9	White	9	Silver	× .01	Silver	± 10%

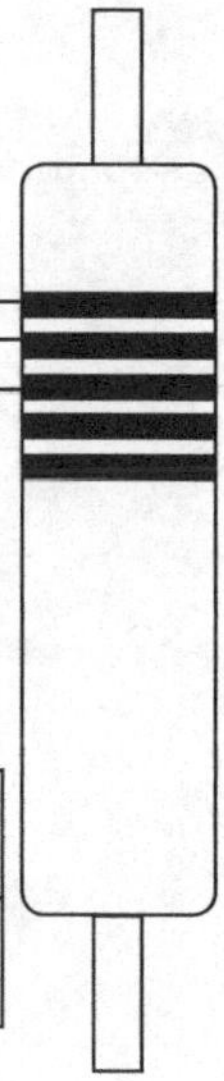

2.6.1 Measurement of Inductance

Like the capacitance test, the inductance measurement is best performed using a 1 kHz energizing supply source. include the RLC bridge, the Maxwell inductance bridge, the Hays bridge, and others. The self inductance of a coil is the basis for the inductance measurement, which is expressed in terms of the (H) Henry. This bridge is built on the idea of comparison, which means that the measured value of the self-inductance is compared to this standard and the unknown inductance value is derived by comparing the value of the unknown resistance.

- L1 = R1's inductance, which is unknown.
- L2: Fixed-resistance variable-inductance coil.
- An inductor L2 with a variable resistance, R2.
- A known non-inductance resistance is R3, R4.

Know more about inductance

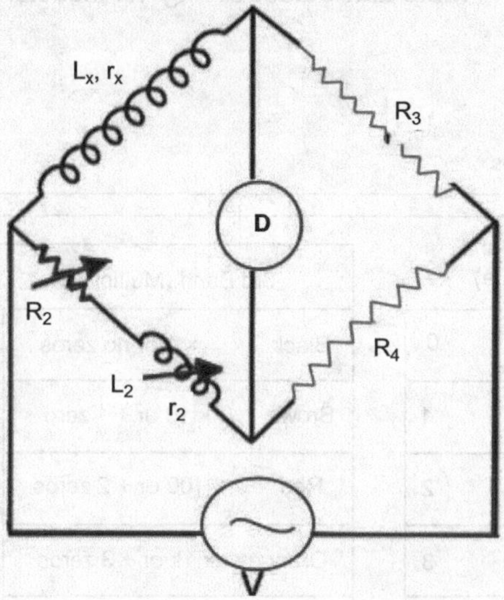

Fig. 2.27: Maxwell's inductance bridge

L_1 = Unknown inductance

$$L_1 = \frac{R_3}{R_4} L_2$$

At balance,

$$R_1 = \frac{R_3}{R_4}(R_2 + r_2)$$

R_1 = Inductor Resistance

Using the resistance box, we can change the value of resistors R3 and R4 from 10 to 1000 ohms. The addition of a resistance into the circuit is required for bridge balancing.

2.6.2 Testing of Inductors

Testing of resistance across its leads is main process of inductance testing,Whatever be the type of the inductor, depending on core or fixed and variable, the resistance of the winding is measured.

Following are the procedure of testing of inductor

- In multimeter select resistance measurement and put dial on lowest range of resistance
- Then connect the probe tips across the inductor terminals
- At this time the multimeter will shows very low resistance because its only winding wire resistance. The resistance value depends upon length and gauge of wire and it is varies from inductor to inductor. but as if wire is of conducting material than resistance should be very low
- If the multimeter shows the "open" it indicates disconnected coil wire or may be end terminal are disconnected with the coil.
- Multimeter "zero" value resistance showing damaged insulation of the coil and it giving rise to lateral electrical connections as well as with the core

- If the core is reachable, the resistance between the core and the terminals should be measured .A good coil should give"infinite" or "open" resistance on the multimeter.
- The same testing should be done with the external metallic body of the choke coil.If it is present

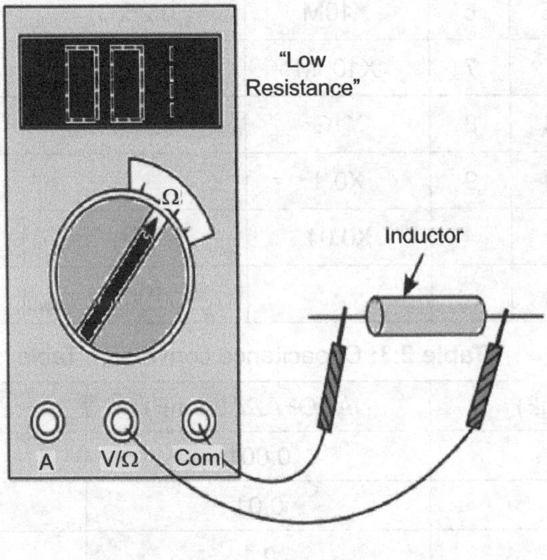

Fig.2.28: Testing of an inductor

- A continuity test with an ohm meter will reveal any open circuits in the inductors. In most cases, a review of the waveform response will reveal any inductors that are shorted or partially shorted.

When high frequency signals are delivered across the circuit, partial shorting often reduces high frequency responsiveness (roll-off).

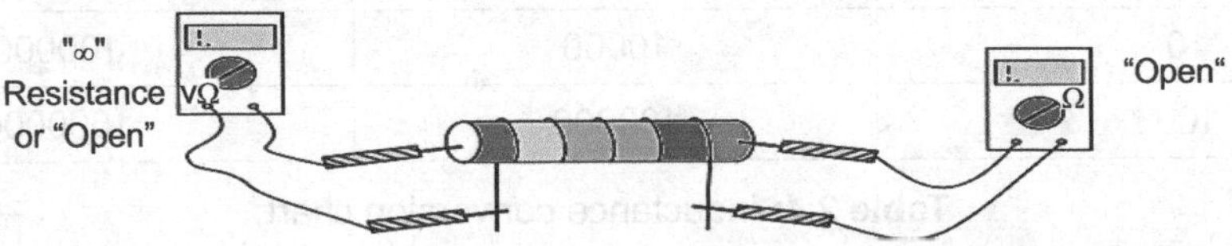

Fig. 2.29: Testing resistance between core and winding of an inductor

Table 2.2: Resistor color code analysis

COLOUR	SIGNIFICANT FIGURES			MULTIPLY	TOLERANCE (%)	TEMP.COEFF. (ppm/K)	FAIL RATE (%)
BLACK	0	0	0	X1		250(U)	
BROWN	1	1	1	X10	1(F)	100(S)	1
RED	2	2	2	X100	2(G)	50(R)	0.1
ORANGE	3	3	3	X1K		15(P)	0.01

YELLOW	4	4	4	X100K		25(Q)	0.001
GREEN	5	5	5	X1M	0.5(D)	20(Z)	
BLUE	6	6	6	X10M	0.25(C)	10(Z)	
VOILET	7	7	7	X100M	0.1(B)	5(M)	
GREY	8	8	8	X1G	0.05(A)	1(K)	
WHITE	9	9	9	X0.1			
GOLD				X0.01	5(J)		
SILVER					10(K)		

Table 2.3: Capacitance conversion Table

MICROFARADS(uF)	NANOFARADS(nF)	PICOFARADS(pF)
0.000001	0.001	1
0.00001	0.01	10
0.0001	0.1	100
0.001	1	1000
0.01	10	10000
0.1	100	100000
1	1000	1000000
10	10000	10000000
100	100000	100000000

Table 2.4: Inductance conversion chart

picohenry (pH)	**nanohenry (nH)**	**microhenry (µH)**
1	0.001	0.000001
10	0.01	0.00001
100	0.1	0.0001
1000	1	0.001
10000	10	0.01
100000	100	0.1
1000000	1000	1
10000000	10000	10
100000000	100000	100

Table 2.5: Resistance conversion chart

Unit Name	Relation to ohm
Abohm(abΩ)	1E-9[Ω]
EMU of resistance	1E-9[Ω]
ESU of resistance	8.987552E+11[Ω]
ohm	1[Ω]
Milliohm	0.001[Ω]
Killoohm	1000[Ω]
Megaohm	1E+6[Ω]
statohm	8.987552E+11[Ω]
Kilostatohm	8.987552E+14[Ω]
Volt/Ampere	1[Ω]
Volt/Kiloampere	0.001[Ω]
Volt/Milliampere	1000[Ω]
Millivolt/Kiloampere	1E-6[Ω]
Millivolt/Megaampere	1E-9[Ω]
Kiloavolt/Milliampere	1E+6[Ω]
Kiloavolt/Ampere	1000[Ω]
Kiloavolt/Kiloampere	1[Ω]
Kiloavolt/Megaampere	0.001[Ω]
Megaavolt/Miliampere	1E+9[Ω]
Megaavolt/Ampere	1E+6[Ω]
Megaavolt/Kiloampere	1000[Ω]

UNIT SUMMARY

- Resistors, inductances, capacitors, and transformers are all examples of passive devices that are essential to the construction of any electrical or electronic circuit.
- Unlike active components, which generate or deliver power to a circuit, passive devices drain power inside that circuit as they work like an attenuator, reducing the signal strength.
- Except for components like electrolytic capacitors that are marked with their polarity, all passive components can be connected in either direction within a circuit.
- Important yet passive components like resistors, capacitors, and inductors must be accurately measured.
- Resistors are devices used in electrical circuits to stop the flow of current; the quantity of current stopped depends on the resistor's resistance and may be changed as required.Resistors can be either

linear or nonlinear, depending on their shape and design, and are further categorized by their resistance.

- Fixed resistors are those whose resistance value does not change. Tolerance, cost, and noise are just some of the resistor attributes that can be impacted by the use of alternative materials.
- A variable resistor is a resistor whose resistance value may be changed. These resistors have a wiping contact and a revolving shaft.
- Nonlinear resistors. are those whose voltage & current characteristics of resistors can be changed linearly like varistors, thermistor, LDR & SMD.
- Testing of resistance can be measure by using an ohmmeter or analog multi meter
- Excessive heat, applied voltage, humidity, mechanical stress, and vibrations all cause chemical or other changes that eventually lead to resistor failure by gradually deteriorating them.
- The colour coding is the method to find resistance value of the resistors and it is done in the form of coloured rings each colour is a code for a particular number, while the location of this coloured ring decides the placement or multiplier of the number.
- Potentiometers are a form of variable resistor that have a track of resistance material that is made electrical contact with through a moveable wiper.
- Inductors are two types fixed and variable types .The construction difference is the movement of core that is movable or fixed.
- Similar to testing fixed resistors, a potentiometer's performance may be evaluated using an ohmmeter; the test should be carried out between the terminal with the variable contact and the two fixed terminals.
- Cadmium sulphide is used to construct LDRs. When stored in total darkness, there are extremely few free electrons for them to interact with, resulting in a very high resistance. Light may free electrons, making a substance more electrically conductive.
- The capacitor is a device that stores energy by accumulating an electrical charge.
- Capacitor can be tested using multi-meter by measuring its resistance and voltage after charging.
- If multi-meter not having increasing value of resistance shows low or approximately zero value that indicating that capacitor is partially or completely short respectively
- The voltage across a capacitor directly propotional to charge,and vice versa.Therefore,we can test the capacity or charge retaining capabilty of capacitor by measuring voltage across a charged capacitor
- A capacitor's dielectric is the insulating layer between its plates; it's what determines whether the capacitor is an electrolytic one, a paper capacitor, a polyester capacitor, a ceramic capacitor, a mica capacitor, or something else entirely.
- In capacitor due to dielectric breakdown ,high voltage,and environment factors ,inter connection of the metal plates,or due to breakage in "in-between"dielectric material or misalignment at the ends capacitor can get short
- The capacitor can get leaky due to degradation of the dielectric properties of the dielectric with aging, over-voltage application,environmental factor like humidity, or substandard quality of the material.

- The capacity of a capacitor can be measured with the help of many RLC meter, Schering bridge etc. The capacitance can be measure by connecting the capacitor in the bridge and balancing the bridge.
- According to the "self inductance" principle, which governs how inductors work, any change in the current running through a wire coil results in the creation of a magnetic field within the inductor.
- Testing of resistance across its leads is main process of inductance testing.
- Measurements of inductors and capacitors should both be taken at the supply source. include the RLC bridge, the Maxwell inductance bridge, the Hays bridge, and others.

EXERCISES

Multiple Choice Question

1. What will be resistor color code 3 band for 220 ohm resistors with 20% tolerance Red, Red, Brown
(a) Brown, Black, Orange (b) Violet, Blue,Gold
(c) None of these (d) Brown, Black, Orange

2. What will be color code of 10K resistors with 20% tolerance
(a) Brown, Black, Orange (b) Red, Orange, Blue
(c) Violet, Blue, Gold (d) None of these

3. 20k ohm resistor color code is
(a) Red, Black, Orange, Golden (b) Red, Blue, Orange, Silver
(c) Yellow, Black, Orange, Silver (d) Red, Orange, Golden, Blue

4. In order to achieve high accuracy, the slide wire of a potentiometer should be
(a) As long as possible (b) As short as possible
(c) Neither too small not too large (d) Very thick

5. To measure an A.C. voltage by using an A.C. potentiometer, it is desirable that the supply for the potentiometer in taken
(a) From a source which is not the same as the unknown voltage
(b) From a battery
(c) From the same source as the unknown voltage
(d) Any of the above

6. The typical tolerance of the paper capacitors is ________
(a) 2% (b) 4%
(c) 6% (d) 10%

7. The advantage of the paper capacitor is ________
(a) Low cost (b) Voltage rating is high
(c) Stable (d) All of the above

8. How many conductors does the capacitor consist of?
(a) One (b) Two
(c) Three (d) Four

9. The capacitance measured in ______
   (a) Watts (b) Henry
   (c) Farads (d) None of the above
10. The dielectric materials increase the ______
   (a) Capacitance (b) Resistivity
   (c) Both a and b (d) None of the above
11. The relationship between frequency, capacitance, and capacitor reactance is given by ______
   (a) Capacitance reactance=1/2πfc (b) Capacitance reactance=1/2πf
   (c) Capacitance reactance=1/2fc (d) Capacitance reactance=1/4πfc
12. Which one is a polarized capacitor?
   (a) Electrolytic capacitor (b) Ceramic capacitor
   (c) Silver capacitor (d) Teflon capacitor
13. The capacitance range of super capacitors is from ______
   (a) 100 farad to 5000 farad (b) 200 farad to 5000 farad
   (c) 300 farad to 5000 farad (d) 400 farad to 5000 farad
14. Basically a potentiometer is a device for
   (a) Comparing two voltages (b) Measuring a current
   (c) Comparing two currents (d) Measuring a voltage
15. The Ohm's law obeys components are known as:
   (a) Ohmic components (b) Non-ohmic components
   (c) Resistors (d) None of these
16. Which of the following inductor will have the least eddy current losses?
   (a) Air core (b) Laminated iron core
   (c) Iron core (d) Powdered iron core
17. A crack in the magnetic path of an inductor will result in
   (a) Unchanged inductance (b) Increased inductance
   (c) Zero inductance (d) Reduced inductance
18. A laminated iron core has reduced Eddy current losses because
   (a) More wire can be used with less DC resistance in coil
   (b) The laminations are insulated from each other
   (c) The magnetic flux is concentrated in the air gap of the core
   (d) The laminations are stacked vertically
19. The coefficient of coupling between two air core coils depends on
   (a) Self inductance of two coils only
   (b) Mutual inductance between two coils only
   (c) Mutual inductance and a self inductance of two coils
   (d) None of the above
20. Both the number of turns and the core length of an inductive coil are double. Its self inductance will be
   (a) Unaffected (b) Doubled
   (c) Halved (d) Quadrupled

ANSWERS															
1.	(a)	2.	(a)	3.	(a)	4.	(a)	5.	(c)	6.	(d)	7.	(d)	8.	(b)
9.	(c)	10.	(a)	11.	(b)	12.	(a)	13.	(a)	14.	(a)	15.	(c)	16.	(a)
17.	(d)	18.	(b)	19.	(c)	20.	(b)								

Short Answer Type Question

Category 1

1. What is resistor explain and how do you calculate resistor color code.
2. What is 2.2k ohm Resistor color code for 4-band.
3. What is colour code for 100K resistor.
4. What does a 1k resistor color coding
5. How is a 3.9 K resistor color code.
6. Calculate a 6-band resistor value.
7. Calculate any 3 bands resistor value.
8. Determine the total inductance in series of a 6 H and a 4 H inductor
9. Determine the total current flow for the circuit if Inductors of 1H and 2H are connected in series to a 440V, 60Hz power supply.
10. A parallel plate capacitor with a capacitance of 4 F can store a 600 C charge. How much of a potential difference exists between the capacitor's plates.
11. Which bridge is more efficiently using for measurement of inductance.Explain?
12. Which is the first band for the colored-resistors?
13. An AC voltage of 240 volts with a frequency 60 Hz is applied to a 0.5 H inductor. neglecting its small amount or wire resistance, how much current would flow through it?
14. State the relationship between the inductance value of a coil and the amount of emf it produces.
15. Define the term inductive reactance.
16. Define paper capacitor.
17. List various applications of ceramic capacitors

Category 2

1. Calculate the RL time constant for a coil that has an inductance of 5H and a DC resistance of 10 Ω.
2. A parallel plate capacitor of capacitance 4μF carries a charge of 600 μC. What is the potential difference between the plates of the capacitor?
3. Find the equivalent capacitance and. the total charge and potential difference across each capacitor is 12V. And the charge on each capacitor if two capacitors C1= 1 micro farad and C2= 2 micro farad are connected in parallel.
4. Analyze the characteristics of passive components?
5. Summarize the different type of thermistors
6. Define basic unit used to measure the resistance of a resistor?
7. Describe the different types of fixed resistors

8. Discuss the most commonly needed variable resistors
9. List various applications of wire wound resistors.
10. Explain the advantages and disadvantages of carbon composition resistors in details.
11. Define metal film resistor and what are the advantages of metal film resistors?
12. Summarize the applications of capacitor in electronics and electrical field.
13. What is the basic construction of a capacitor?
14. What is the total capacitance of the combination and potential difference across each capacitor if the combination is connected to a 120 V supply and four capacitors each of capacitance 3 pF are connected in series.
15. How many categories of problems arise due to passive component failure.

REFERENCES AND SUGGESTED READINGS

1. Cooper, W.D. (1984), Electronic Instrumentation and Measurement Techniques, Prentice Hall of India Private Ltd., New Delhi. ISBN 0-87692-060-1
2. Type 5MC Metallized Polycarbonate Capacitor, Electronic Concepts, Inc., Box 1278, Eatontown, NJ, 07724, (908) 542-7880.
3. " Troubleshooting of Electronic Equipment " R.S khandpur book was first pubJished in indiai 2003by Tata McGmw-Hill.
4. Tantalum Electrolytic Capacitor SPICE Models, Kemet Electronics, Box 5928, Greenville, SC, 29606, (803) 963-6300.
5. Electrical Components: A Complete Reference for Project Builders October 1991 by Delton Horn McGraw-Hill edition
6. https://www.sciencedirect.com/science/article/pii/B9780081017647000037
7. https://nptel.ac.in/courses/122106026
8. https://archive.nptel.ac.in/content/storage2/courses/122104013/node4.html
9. https://nptel.ac.in/courses/115106122

KNOW MORE

An incandescent bulb that may be regarded passive element may serve as a current stabilizer. It may be used as Automatic Gain Control(AGC) element too. and an electrolytic capacitor(passive element) is semi-conductive and thus may amend interspersing current the same as a diode.

QR SCAN CODE FOR SUPPORTIVE KNOWLEDGE EMBEDDED IN THE CHAPTER

□□

3 Testing of Semiconductor

UNIT SPECIFICS

The following topics have been addressed in this unit:

- Understanding the types of semiconductor devices
- Failure analysis of semiconductor devices: understanding the root causes
- Types ,applications and testing procedures for diodes, thyristors .
- Working principle and Fault diagnosis in op-amp circuits.

The study of these issues' real-world applications is meant to inspire more inquiry and originality, as well as to sharpen the ability to solve problems. The unit includes knowlege of electronic device troubleshooting problems, a list of references, and suggested readings that can be used for practise, as well as a large number of multiple choice questions and questions with short and long answers divided into two categories based on the lower and higher orders of Bloom's taxonomy. It's worth noting that QR codes, which may be scanned for further information, have been included in various parts so that readers can learn more about the themes that interest them.

There is a "Know More" section depending on the information that comes after the associated practical. The supplemental material included here has been carefully organized to maximize its usefulness to readers. This section focuses on the beginning activity, examples of some fascinating facts, analogies, the history of the subject's evolution while emphasizing significant observations and findings, timelines beginning with the creation of the relevant subjects up to the present, applications of the subject matter for our day-to-day real life or/and industrial applications on a variety of aspects, and case studies related to environmental, sustainability, social, and ethical issues.

RATIONALE

This unit on testing of semiconductor devices helps students to get basic idea about the semiconductor devices etc . It explains the variety of semiconductor devices and their limitations and applications in daily life. This unit will also explain various concept and correct methodology of testing semiconductor devices .And it would reduce the chances of failure.

PRE -REQUISITES

Physics: *Semiconductor Electronic: Material, Devices and Simple Circuits (Class XII)*

Moving Charges and Magnetism (Class XII)

UNIT OUTCOMES

List of outcomes of this unit is as follows:

U3-O1: Understand the working principle of semiconductor devices

U3-O2: Describe Thyristors or Silicon controlled rectifier and its applications

U3-O3: Analysis and diagnose the types of failure in semiconductor devices

U3-O4: Explain the differences between bipolar junction transistors and field effect transistors.

U3-O5: Describe that op-amp circuits operate, use amplifiers, and identify faults.

Unit-3 Outcomes	EXPECTED MAPPING WITH COURSE OUTCOMES (1 - Weak Correlation; 2 - Medium Correlation; 3 - Strong Correlation)					
	CO-1	CO-2	CO-3	CO-4	CO-5	CO-6
U3-O1	1	3	3	-	3	1
U3-O2	1	2	2	2	2	1
U3-O3	2	1	1	2	3	3
U3-O4	2	1	1	1	2	1
U3-O5	3	2	2	1	3	2

3.1 INTRODUCTION OF SEMICONDUCTOR DEVICES

Semiconductor devices are made up of materials that fall somewhere in the middle of the conductivity spectrum, between a good conductor and an insulator. Pure elements like silicon and germanium are examples of semiconductors, but materials like gallium arsenide and cadmium selenide, which have had their conductivity altered by the addition of small quantities of impurities via a process called doping, are other examples. The construction of the electrical devices we use on a daily basis depends mainly on semiconductors. Because to their robustness, portability, and low cost, and they have a broad variety of applications. These semiconductors can endure a broad range of currents and voltages and are used in power electronics, optical sensors, and light emitters. Their ratings for voltage and current are over 5,000 amps and 100,000 volts, we can say that semiconductor is a key element of electronic devices, enable advancements in communication, computing, healthcare, military systems, transportation, renewable energy, and many other fields. You may find them in high-tech gadgets including smartphones, radios, televisions, computers, video games, and medical diagnostic tools. The ability of semiconductors to regulate the flow of electrons makes them useful in a wide variety of electronic circuits. In addition to their usage in electronics, semiconductors find use in fields such as solar energy harvesting due to their sensitivity to light. The solar cells' have ability to produce electricity that is in proportional to the quantity of light incident onto the semiconductors. The conductance increases when there are more free electrons accessible, which is when an N-type semiconductor is used. A P-type semiconductor is employed when the inductance is higher yet there are fewer free electrons. The resistance of semiconductor materials is higher than that of an insulator but lower than that of a conductor. Due to the negative temperature coefficient of resistance of the materials, the resistance of semiconductors decreases as their temperature increases. A semiconductor device has a far greater chance of malfunctioning than a passive component. They are having more probabilities to failure in comparison to capacitor or resistor etc. because semiconductor more sensitive to heat. They can easily damaged due to excess heat and temperature. Due to their small size of device results in reduction of its surface area and lessor will be the heat reduction happens due to small surface area. And these small devices are unable to radiate heat at same rate as it is developed in them. This result in continuous rise in its temperature which can burn or damage the device.

3.1.1 Types of Semiconductor Devices

Semiconductor devices play an important role in electronic equipment, such as their usage as variable resistors, rectifiers, amplifiers, oscillators, modulators, voltage and current sources, electronic switches, and voltage shifters. They can be used as an energy converter (like an LED) and to generate logarithmic and anti-logarithmic functions. There are two primary kinds of semiconductor devices:

Bi-polar : For a bipolar device to function, charge carriers must move freely across both the forward and reverse bias junctions. The current through the collector base junction of a NPN bipolar transistor is governed by the forward biased base emitter junction when the base is reverse biased. Some of the electrons that make it to the base collide with holes, but the vast majority of them disperse and get picked up by the collector. Transistors, diodes, uni-junction transistors, thyristors, TTL and linear integrated circuits, and other similar components are all examples of bipolar devices.

Uni-polar: The electrostatic field between the gate and the source or the gate and the substrate is used to regulate current flow in uni-polar devices, which make use of only the majority carriers. Linear integrated circuits include the Field Effect Transistor (FET), Metal Oxide Semiconductor Field Effect Transistor (MOSFET), complementary metal oxide semiconductor (CMOS) logic, and many others.

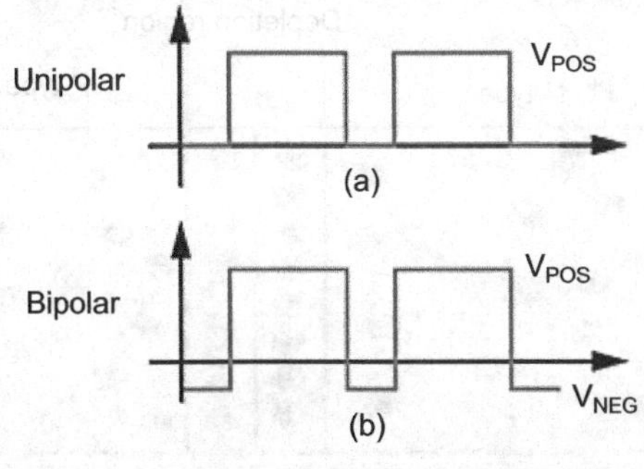

Fig 3.1: Representation of Uni-polar and Bi-polar

Negative charge carriers : Transporting a negative charge from one location to another requires the use of charge carriers, such as free electrons. When an electron loses its bond with its parent atom, it becomes a free electron, which may then travel independently of the atom.

Positive charge carriers : In order to transport positive charge from one location to another, positive charge carriers are used. Holes are valence band voids that migrate from one level to another.

Majority and minority charge carriers definition : A system's preponderant charge carriers are known as its "majority charge carriers." The vast majority of a semiconductor's charge or current is transported by its majority charge carriers. As a result, the majority charge carriers in a semiconductor are primarily responsible for the conduction of electricity. It's possible to have a small number of minority charge carriers. Minor charge carriers in semiconductors carry fewer electric charges or currents..

Charge carriers in intrinsic semiconductor : When the number of free electrons (negative charge carriers) is precisely equal to the number of positive charge carriers, a semiconductor is said to be intrinsically semiconductor (holes or vacancy).

Majority and minority charge carriers in n-type semiconductor : An n type semiconductor is created when the intrinsic semiconductor is bombarded with pentavalent elements like phosphorus or arsenic. There is a plethora of unbound electrons in N-type semiconductors. Therefore, free electrons make up the vast majority of charge carriers in an n-type semiconductor. In an n-type semiconductor, the majority charge carriers, or free electrons, are responsible for the vast majority of the charge and current. The presence of holes is extremely rare in n-type semiconductors. In an n-type semiconductor, holes are the underdog, as they make up a small fraction of the total charge carriers. An n-type semiconductor's holes (minor charge carriers) don't transport much juice. In an n-type semiconductor, the positive charge carriers are more numerous than the negative charge carriers (free electrons) (holes).

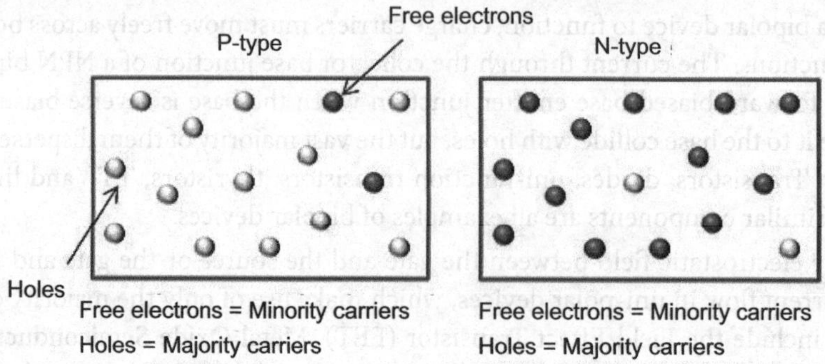

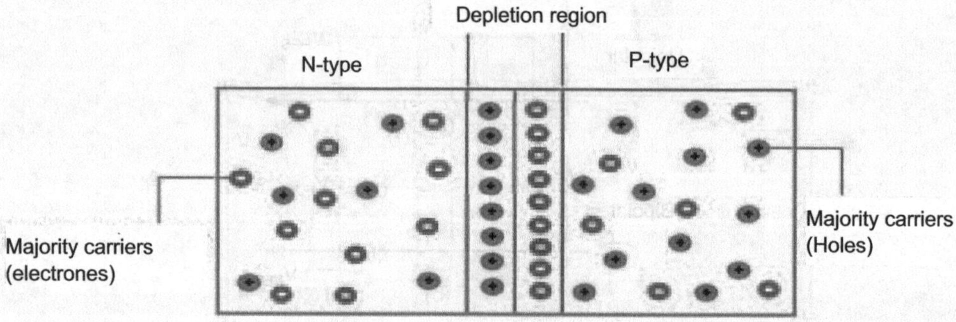

Fig. 3.2: Flow of charge type carriers

Majority and minority charge carriers in p-type semiconductor : If you combine an intrinsic semiconductor with a trivalent element, like gallium or boron, you get a p type semiconductor. Holes abound in P-type semiconductors. This means that holes predominate as charge carriers in p-type semiconductors. To put it another way, the holes in a p-type semiconductor are the primary carriers of electric charge and current. There is a scarcity of unpaired electrons in P-type semiconductors. This means that in a p-type semiconductor, free electrons are a rare occurrence. In a p-type semiconductor, free electrons make up the minority of charge carriers but can only transport a negligible amount of electricity. Holes, the positive form of charge carrier, are more scarce in P-type semiconductors than their negative counterparts (free electrons).

3.1.2 Causes of failure in Semiconductor Devices

Any of the following may lead to a failure in semiconductor devices:

- The gadget may have a flaw in its construction that causes it to malfunction before its time. The problems might be caused by a short circuit or an open circuit.
- Problems may arise from improper use or from careless construction or testing
- The failure of a semiconductor device can occur if its voltage, current, or power levels are increased above its acceptable operating range.
- One of the leading reasons of early failure in semiconductor devices is electrical interference. Semiconductor junction failure is commonly triggered by voltage spikes sent down the mains leads due to the switching of heavy machinery or relays.

3.1.3 Types of Failure Semiconductor Device

Most electrical breakdowns happen because of a connection that is open or has a short circuit. In a bipolar transistor, a short or open circuit might emerge between the base and emitter or the collector and base. There is also a chance of emitter-collector short circuits.The common failure mechanisms for semiconductors due to some of following categories

- Encapsulation failure. If the device's casing develops a crack, then the fault is manifested. Causes of these cracks could include mechanical or thermal stress, as well as variations in the coefficient of thermal expansion between the encasing material and the metal used for the leads. If the relative humidity is high or if the device is exposed to flux, cleaning agents, etc., moisture may enter the packing through these perforations. The semiconductor device could be harmed by chemical reactions.
- Die-attach failure. The thermal conductivity between the die and substrate is impaired by poor contact. This may cause the die to overheat, which in turn stresses and cracks the die, eventually resulting in a broken gadget.
- Wire bond failure. Wire-bond failures can occur for many different reasons, including overheating from excessive current flow, mechanical stress in the bond wire due to insufficient bonding, bond wire and die contact fractures, electromigration of silicon, and too much bonding pressure. One of the gadget›s conductors has been severed, causing the bond to fail.
- Bulk silicon defects. Device malfunctions might be brought on by crystal defects, impurities, and contaminants in silicon bulk material. Failure of the device may also be the consequence of process flaws, such as those brought on by fabrication-related diffusion problems.
- Oxide layer faults. Thin oxide coatings act as insulators and may fail due to electrostatic discharge or high-voltage transients that travel through the leads of an electronic device. Faults may also be caused by the presence of cracks, scratches, or contaminants in the oxide layer
- Aluminum metal faults. These flaws are a result of aluminium corrosion, metal wear from soldering, inappropriate metal deposition at contact windows, hillock growth, and cracking. A failure often doesn›t happen until after a certain event or set of conditions has happened.

- Thermal over stress Damage to semiconductors may be caused by an excessive amount of heat. Materials melt, polymers burn, semiconductor dies stretch and crack, and other damage occurs when temperatures go too high. Junction temperatures shouldn't get higher than 125-150 degrees Celsius.
- Electrical over stress

  Manufacturer specifications for voltage, current, and power levels should be respected while using semiconductor devices. Power and input/output (I/O) connections to a device are subject to the same constraints. Operating a device beyond of its "safe operating area" (SOA) exposes it to the risk of electrical over stress (EOS), which may lead to an internal voltage breakdown and, eventually, damage or destruction

3.2 DIODES

Diodes are electrical devices in which electric current flows in only one direction, and their cathode and anode ends are clearly marked. Identifying markings are symbols or marks on the body's surface that may be used for identification purposes. Stripes, multiple stripes, or a dot are commonly used to mark the cathode end of diodes housed in glass. A silicon or germanium diode's colour will typically correspond to the manufacturer's component number. Figure 3.3 displays symbol of diode.

Measuring the forward and reverse resistance of a diode with an ohmmeter is a quick and easy way to check the device's functionality. The resistance of a standard signal diode or rectifier, in one direction, should be extremely small (a few hundred ohms), while the resistance in the other way, should be very large (near infinity). The resistance reading shouldn't be close to 0 ohms (shorted) or open in both directions. The following considerations are important when using an ohmmeter to test a diode

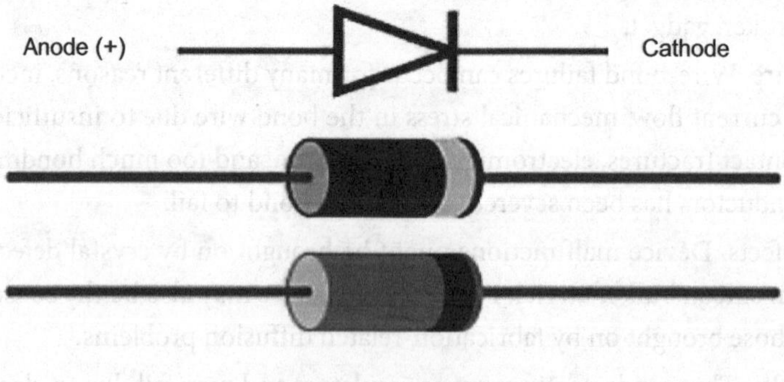

Fig. 3.3: Diode symbol

Diodes may be examined using a wide range of ohmmeters, each of which will provide a seemingly different reading for the forward resistance. Likewise, various ohmmeter ranges will expose you to forward resistances of varying levels. Diode's non-linear (Fig.3.4) voltage or current characteristics cause this change in resistance.

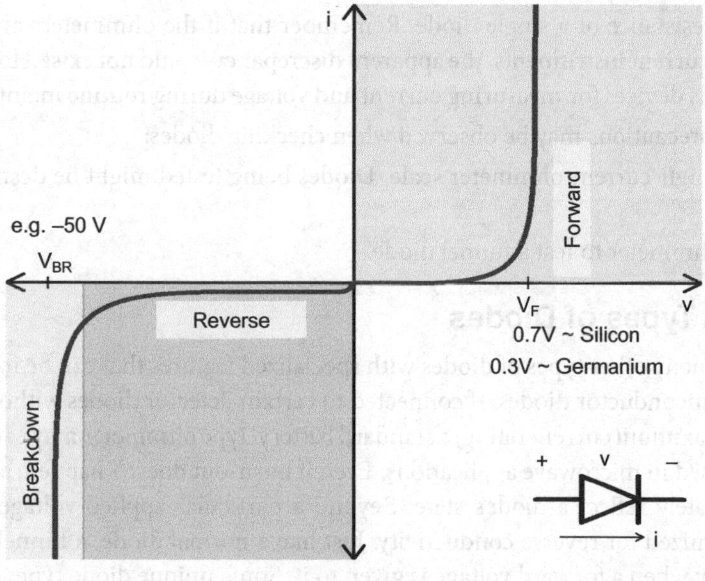

Fig. 3.4: Voltage current characteristics of semiconductor devices

Contrary to appearances, multiple ohmmeters, even the same ohmmeter in different ranges, may not utilize the same test voltage.

- Before testing a diode with an ohmmeter, you should always remove one end from the circuit, as even small amounts of shunt resistance (if present in the circuit) can cause huge distortions in on-the-spot readings.
- A germanium diode has substantially lower forward resistance than a silicon diode, but the reverse resistance of a silicon diode is often measured to be much greater.
- The diode test mode is often found on digital (DMM) multi-meters. Using this technique, a silicon diode should provide a forward voltage reading of between 0.5 and 0.8V and a reverse voltage reading of open. Forward voltage dips on germanium diodes range from 0.2 to 0.4 V. Conventional resistance ranges cannot determine the forward drop of a semiconductor junction because the metre does not apply a voltage high enough to do so. However, it's crucial to keep in mind that a defective diode, especially at the maximum ohms, may exhibit a resistance that is less than infinite. A malfunctioning piece of equipment would thus be indicated by any such readout, albeit this is not always the case.

Sometimes, you need to check the V_f (forward voltage drop) and R_{BR} (Reverse bias resistance) values to see if they're within acceptable ranges (reverse breakdown voltage). You can see the typical behaviour of a diode in Figure 3.3. The peak inverted voltage rating of a stack of diodes is simply the rating of a single diode multiplied by the number of diodes in the stack, making diode stacks a common component in high voltage rectifier circuits. When two diodes are stacked and connected in series, the resulting resistance is greater than twice that of a single diode. If you connect three diodes in series, their resistance will be three times that of a single diode. This discrepancy is again due to the diode's non-linear behaviour. By connecting diodes in parallel, a higher maximum current can be achieved. A parallel connection of two diodes has a current capacity that is twice that of a single diode. Due to the aforementioned causes, it would appear that the resistance of a collection of diodes connected in parallel is greater than the resistance of a single diode. Remember that if the ohmmeters are in fact constant-voltage or constant-current instruments, the apparent discrepancy would not exist .However, ohmmeters are often employed as devices for measuring current and voltage during routine maintenance and repair.

The following precautions may be observed when checking diodes:

- Avoid using a high-current ohmmeter scale. Diodes being tested might be destroyed by excessive currents.
- Never use an ohmmeter to test a tunnel diode.

3.2.1 Special Types of Diodes

However, there are many other types of diodes with specialized features that can be found outside of the general-purpose semiconductor diodes. If connected to certain detector diodes without first identifying their purpose and maximum current rating, a standard battery-type ohmmeter may easily burn them out, particularly those used in microwave applications. Even if burn-out doesn't happen, a simple ohmmeter test may not adequately reflect a diode's state. Beyond a particular applied voltage, the zener diode, for instance, is optimized for reverse conductivity. Just like a normal diode, a tunnel diode's resistance changes dramatically when a forward voltage is given to it. Some unique diode types and their defining features are discussed below.

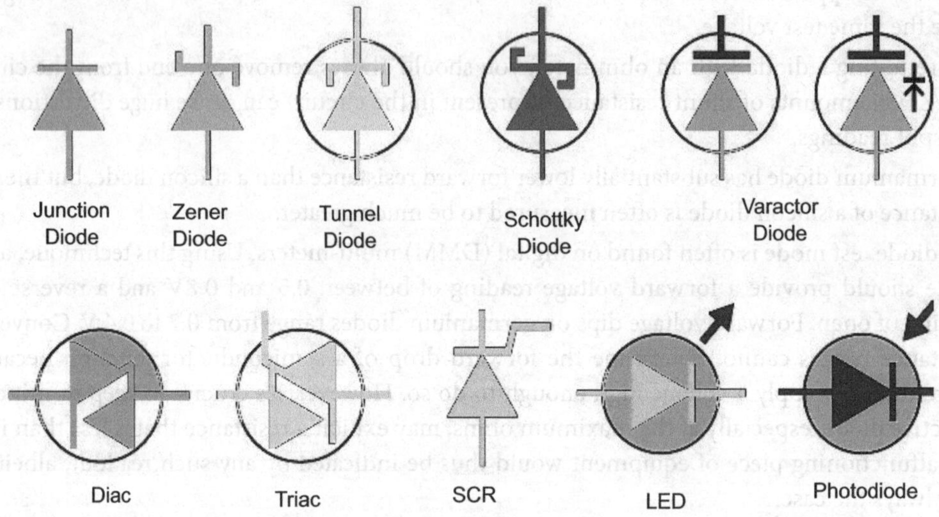

Fig.3.5: Types of Diode

- Zener Diode: The reverse current through a silicon diode is negligible at 25 °C (room temperature). However, once the reverse voltage reaches a certain level, the reverse current increases at an exponential rate. This potential, known as a breakdown avalanche, can be anywhere from a few volts to a few hundred, depending on the materials used in their construction. At bias potentials below the zener voltage, the resistance of a zener diode is extremely high. There could be a few megaohms worth of resistance here. Zener voltage rapidly lowers a zener diode›s resistance to between 5 and 100 Q. When operating in the ‹›zener›› region, where internal resistance is very low, a zener diode can be used as a constant voltage source. If R is a low enough value, the zener current through the diode won›t exceed its rating (Fig. 3.6). Setting the series resistance to a specific amount protects the zener diode›s power rating

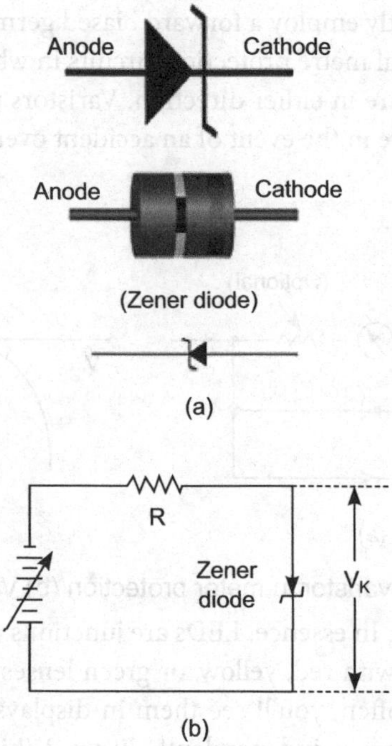

Fig. 3.6: Zener diode: (a) symbol (b) Zener diode as constant voltage source

Application of zener diode

- They are mostly used as 'voltage references' and as 'shunt regulators' to regulate the voltage across small circuits.
- When connected in reverse biased, a Zener diode conducts and produce reverse breakdown voltage
- Varactor Diode: Varactors are silicon diodes that react to varying degrees of reverse voltage in a manner similar to a variable capacitor. Varactors are available for purchase with nominal capacitance values ranging from 1 to 500 pF and maximum rated operating voltages ranging from 10 to 100 V. In automated frequency control circuits, they are often used. A typical varactor has a reverse capacitance of 10 pF at 5 volts and a forward capacitance of 5 pF at 30 volts. Varactor diodes come in a variety of forms, which are seen in Figure 3.7.

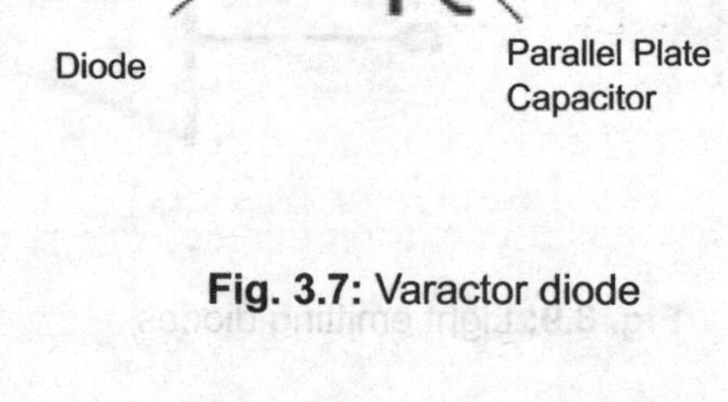

Fig. 3.7: Varactor diode

- Varistor : The resistance of a varistor, a kind of semiconductor, decreases with increasing voltage because it is non-linear in response to voltage. For these reasons, varistor applications, like bias stabilization circuits, frequently employ a forward biased germanium diode. Figures 3.8(a) and (b) illustrate a pair of symmetrical metre protection circuits in which the varistors allow the diodes to route current around the metre in either direction. Varistors prevent potentially harmful voltages from forming across the metre in the event of an accident overload.

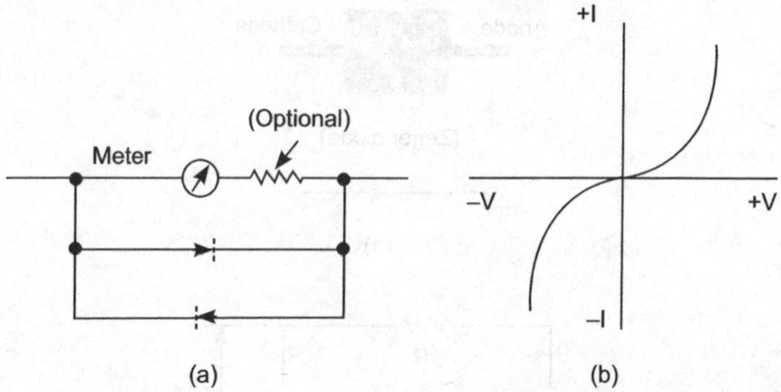

Fig. 3.8: (a) Use of varistor in meter protection (b) Varistor characteristics

- Light Emitting Diodes (LED): In essence, LEDs are junctions that emit light when biased forward. LED lights may be obtained with red, yellow, or green lenses and come in a broad range of sizes and designs (Fig. 3.9).Most often, you'll see them in displays with seven-segment alphanumeric characters, where each segment is independently lit up. While LED displays are used in devices like digital thermometers, calculators, and test equipment, LED arrays are often reserved for more specialized uses like luminance or data collection. LEDs are similar to regular diodes from an electrical perspective, with the exception that they have a larger forward voltage drop. The following are typical instances of these parameters: 1.2 V for 1R (infra-red), 1.85 V for Red, 2 V for Yellow, and 2.15 V for Green. Furthermore, the actual voltages may vary significantly based on the technology used in the LEDs. As a result, the colour of several samples of the same kind of LEDs should be relatively similar, but the LED voltage drop is not a reliable indicator of colour. Therefore, a simple brief and open test using a multi-meter is all that is required to ensure the LED is working properly. However, even if an LED passes electrical inspection, it could not be very powerful. This means that it's important to make sure the LEDs really work. Assuming the other components and connections in the circuit are functional, a faulty LED will be immediately apparent from its behavior in the circuit.

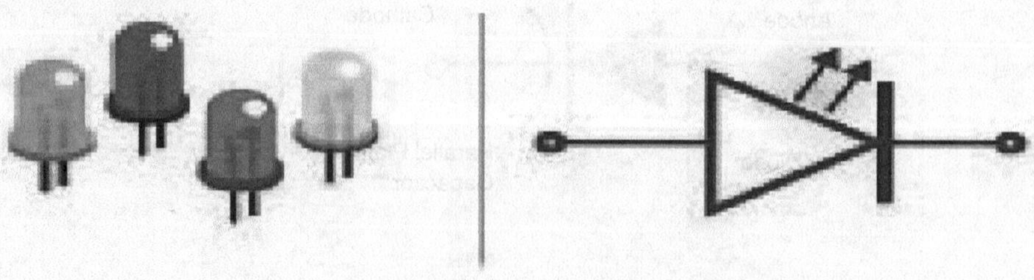

Fig. 3.9: Light emitting diodes

Photo-diode: A solid-state device known as a photo-diode functions similarly to a standard diode but conducts when exposed to light (din junction). It basically functions as an open circuit in the dark, but when illuminated, it conducts a significant amount of energy. When a photo diode is alternately revealed and covered, the voltage drop across a load may be used to assess the diode's performance (Fig. 3.10). (using a dark piece of paper to cover the diode, etc.) The component may switch at the designated current level and should operate properly in the circuit, even if the screen does not show the switching time.

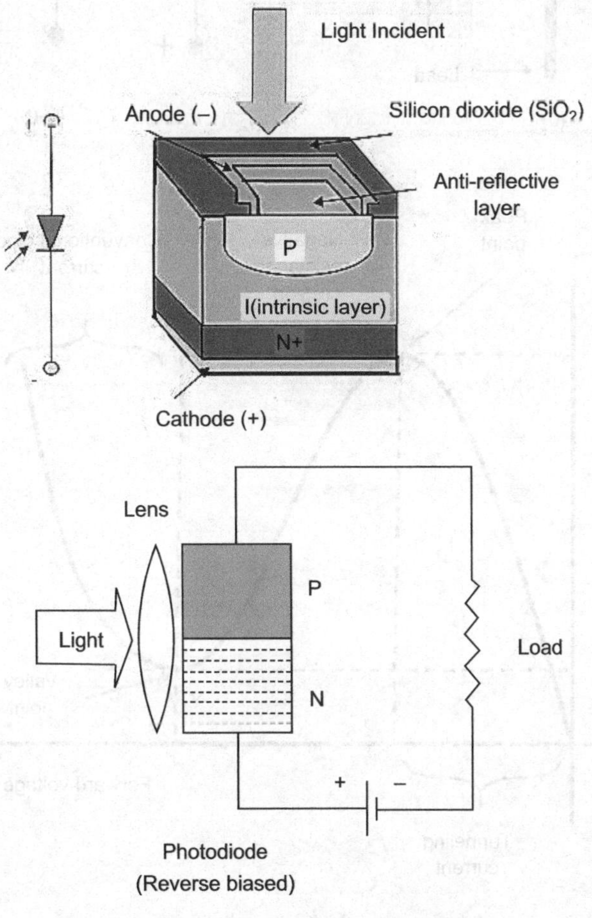

Fig. 3.10: Photo diode structure

Tunnel Diode (TD): A tunnel diode is a pn junction with a negative resistance gap. (Fig.3.11) shows the voltage current characteristics of a tunnel diode. Negative resistance measurements for various types of tunnel diodes range from 1 to 200 ohms. Tunnel diodes may be used to create toggle switches. A switching circuit may be driven from its low current to its high current quiescent point by applying pulses. It is usually beneficial to identify the problem's root cause by rapidly evaluating the tunnel diode's switch-ability and the current level at which it switch

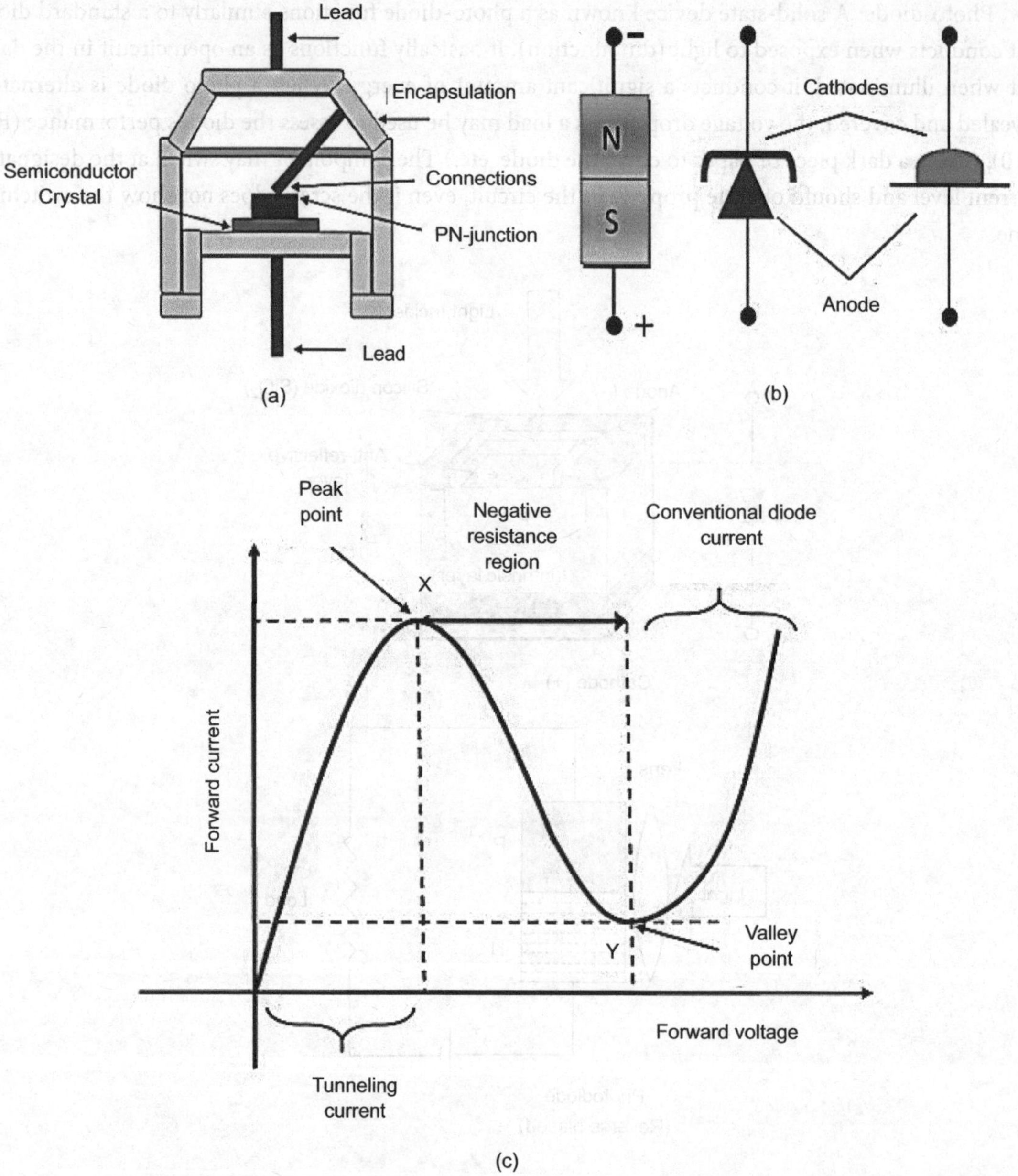

Fig. 3.11: Tunnel diode: (a) Construction (b) symbol (c) V-I Characteristics

Schottky diode: It is a metal semiconductor junction diode. Schottky diode also known as Barrier diode and low voltage diodes , Schottky diode having lower Power drop in comparison to a PN junction diode this diode name based upon 'Walter. H. Schottky' because he is first person who discovered schottky diode

Fig. 3.12: Schottky diode

Applications of Schottky Diode

- Following are the applications of Schottky diodes
- In electronics industry for many applications
- In diode rectifier because of its properties.
- For voltage clamping applications
- To prevent excess transistor saturation.
- As Schottky TTL devices as these devices needs fast switching.
- In digital computers.

3.3 TEST PROCEDURE FOR DIODE

Diodes are one of the most fundamental and essential building blocks of electronic circuits, where they serve a wide variety of functions including protection, rectification, and switching. In the event of a malfunction, they are among the first parts to be harmed. The diode's Anode and Cathode must be determined before the device can be tested. The terminal closest to the white band on the body of most PN Junction diodes is the cathode. The last one is the anode. This symbol may be seen on Diodes with either a through-hole or a surface-mount terminal. The terminal that is closest to the colour of a diode's band, even if it is a different hue, is often the cathode. A diode may be tested in a variety of methods while still adhering to certain fundamental testing guidelines.

3.3.1 Test Procedures for P-N Junction Diodes

The simplest and most fundamental kind of semiconductor device is the PN junction diode. A regular or standard diode is a junction of p- and n-type semiconductors. These days, semiconductors are often made out of silicon rather than Germanium. A diode is an electrical device that only enables electricity to travel in one direction. (Fig.3.3) shows the schematic and symbol for a diode .

A diode may conduct electricity when forward biased, but not when reverse biased. When the device is forward biased, the resistance is low, but when it is reverse biased, it is high.

In other words, it has a low resistance in the direction of the forward bias and an infinite or almost infinite resistance in the direction of the reverse bias. It is rated according to how well it can provide divergent amounts of resistance moving in different directions..

Following are the steps for testing of a pn-junction diode :

- Select the lowest resistance range in the multi-meter. Connect the leads to "V/Ω" and "COM" sockets. Red to "V/Ω" and black to "COM".
- Connect red leads to 'p' side of diode (one without ring) and black lead to 'n' side (one with ring)
- The multi-meter should show a low resistance.This resistance can vary from tens of ohms to few hundred of ohms depending on the voltage in the leads. This resistance will be further less for a Germanium diode.
- Next red lead connect to 'n' side and black lead to 'p' side. The diode becomes reverse biased . It should indicate open-circuit even in maximum range.
- If the diode displays low resistance in both instructions is "short". And if it shows"open" in both instructions is open. If the forward biased resistance is relatively low and reverse biased resistance is somewhat less than infinity, then suggests diode is faulty.

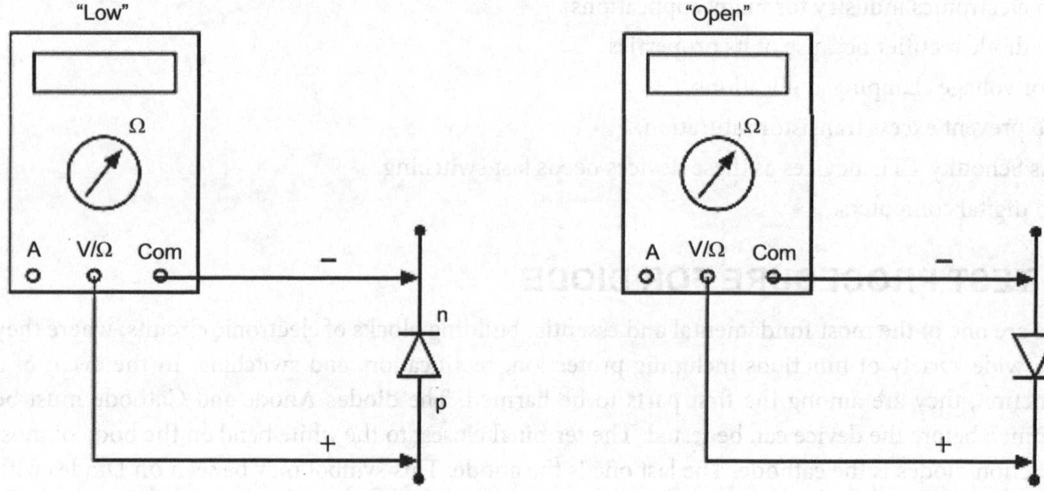

(b)

Fig. 3.13: Testing of diode

Some digital thermometer have diode symbol means that multi-meter having diode checking feature . If the multi-meter is set to the diode symbol position, the display value shows "open" in reverse bias and 0.5 V to 0.8 V in forward bias. With the voltmeter in this setting, it is possible to measure the forward voltage drop across the diode. The 0.5 V to 0.8 V range may be explained by the voltage differential between the two leads. Being a non-ohmic device, the diode resistance is truly voltage dependant.

A Germanium diode's forward voltage drop varies from 0.2 V to 0.4 V. This is a result of the "Germanium" material's low break over voltage. Diodes can also be tested for forward voltage drop by applying appropriate DC voltage. The circuit arrangement requires a resistor, DC supply along with a diode to be tested. The method is as shown in (Fig.3.13) In this procedure, the forward voltage drop and reverse voltage is measured across the diode. When the diode is forward biased ,the multi-meter should display 0.6 V to 0.8 V, or commonly 0.7 across a silicon diode and 0.3 V for a Germanium Diode. These are their forward breaking over voltages as illustrated in Fig.(3.13(a))

The diode ought to behave as an open circuit when placed in reverse bias. As a result, the multi-meter should display the entire input power as shown in (Fig. 3.14). However, keep in mind that the diode's reverse breakdown voltage must not be exceeded when applying a reverse voltage. Since the diode won't conduct current in this configuration, the voltmeter will read the voltage at which the diode undergoes reverse breakdown. This method of diode testing is followed only if the diode testing feature is not available in the multi-meter.

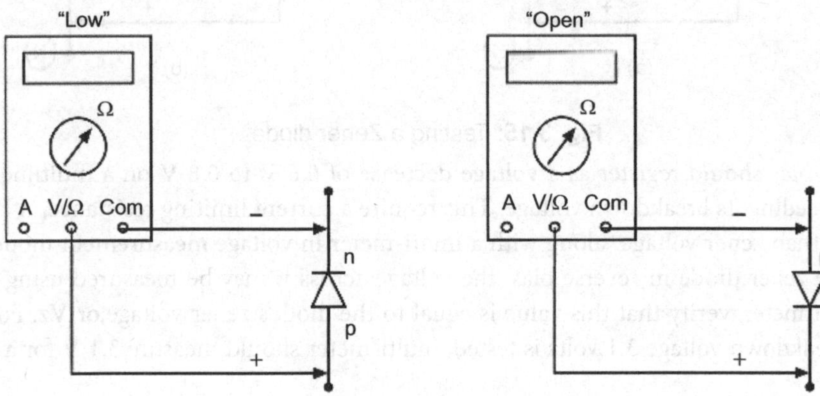

(b)

Fig. 3.14: Testing of forward voltage drop

3.3.2 Test Procedures for Zener Diode

A zener diode's breakdown voltage in reverse needs to be sharp. The breakdown voltage of a zener diode is defined by the device's reverse voltage rating. Because of their analog-like behaviour when reverse biassed, Zener diodes find widespread application as voltage regulators. Below the reverse breakdown voltage, a zener diode's reverse current is zero, but above this voltage, a significant current flows, which is controlled by the external series resistance. External resistance should be select carefully,because any lower value of resistance gives large current through the diode, that current will be more than its current rating and this can easily damage diode. Testing a zener diode with a multi-meter is just like measuring a normal diode. It involves measuring its forward and reverse bias resistance Fig.(3.13)

The zener should show low resistance like regular diode when forward biased,and zener should show "open" in reverse direction. But this occurs if the voltage in the multi-meter leads voltage is larger than the zener breakdown voltage, the multi-meter will read poor resistance in reverse direction likewise. And the user could think that the diode is "short" ,when fact it may not be.Therefore, for correct measurement, the breakdown voltage of the diode should be known as well as the voltage appearing in multi-meter leads. Zener can also be tested by multi-meter in diode testing mode .

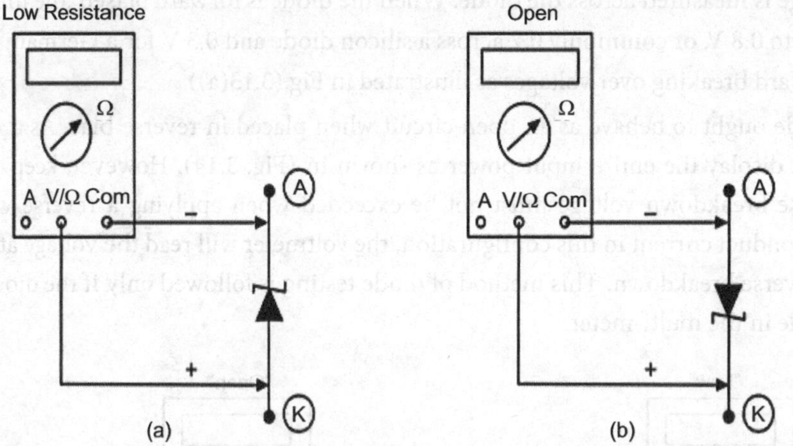

Fig. 3.15: Testing a Zener diode

A forward bias should register as a voltage decrease of 0.6 V to 0.8 V on a multimeter. Under a reverse bias exceeding its breakdown voltage. This require a current limiting resistance, A DC supply of voltage greater than zener voltage along with a multi-meter in voltage measurement mode, (Fig.3.15). Connecting the zener diode in reverse bias, the voltage across it may be measured using a voltmeter. By using a multimeter, verify that this value is equal to the diode's zener voltage,or Vz. For example,if the zener of breakdown voltage 3.1 volts is tested, multi-meter should measure 3.1 V for a good diode. (fig.3.16)

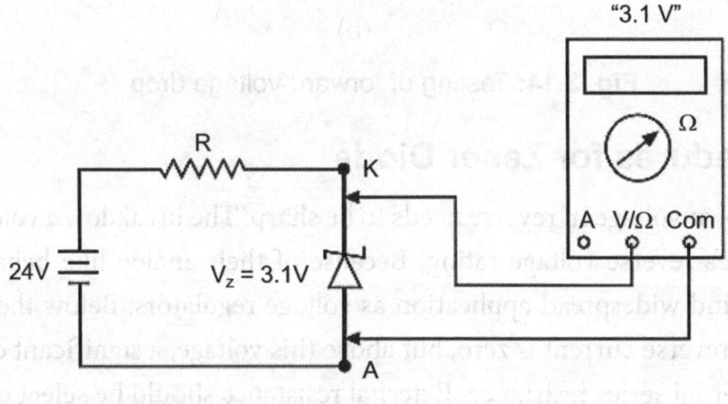

Fig. 3.16: Zener diode showing Breakdown Voltage

If the multi-meter shows the applied Dc voltage as the measurement,then,it means that the diode is open. If the multi-meter shows zero voltage,that means diode is short. So we can say that fault occurring in zener diode are either it is short -circuited or open circuited in both the directions. There is no possibility in shift in zener breakdown voltage. Had it been so, it would have been a "partial failure" for the zener diode.

Precautions

- Its maximum current ratings are not crossed in reverse direction
- At the point when the reverse bias voltage exceeds the zener voltage, the current flows at a very high rate.

- When going to perform the in-circuit testing described above, it is important to select a series resistance that keeps the current below the diode's maximum limit.
- While measuring the voltage in leads is greater than reverse breakdown voltage, There are chances that large current through the zener can damage it.

3.3.3 Test Procedures for Light Emitting Diode (LED)

LEDs are made of semiconductor material, and they produce light when their bias is applied in the forward direction since this is when electron-hole collaboration system. LED are widely used as indicators, in decorative lights etc .Testing a diode is just like testing a normal diode the only difference in both testing is LED's have higher forward voltage drops or higher forward resistance than a regular diode. LED test procedure is as follows:

(a) Select a lowest resistance range of the multi-meter.

(b) Connect the multimeter red lead to the anode (the longer lead) and the black lead to the cathode to put the LED into forward bias (the shorter lead).

(c) The multi-meter should read low resistance. This resistance will be higher than that of a regular diode. If the multi-meter is showing open then switch to next higher range. This is because the voltage in the multi-meter leads may not be enough to forward bias the LED .At higher range, this voltage in starts glowing .

(d) It is possible to change the diode's bias by connecting the red lead to the cathode and the black lead to the anode. The multimeter ought to display zero resistance or virtually infinite resistance at all ranges (small current may flow due to thermally generated carriers in depletion layer)

(e) If the multi-meter has diode testing mode facility, then during forward biasing ,a good LED should show a forward voltage drop ranging from 1.5 volts to 3 volts. And during reverse biasing, it should show open

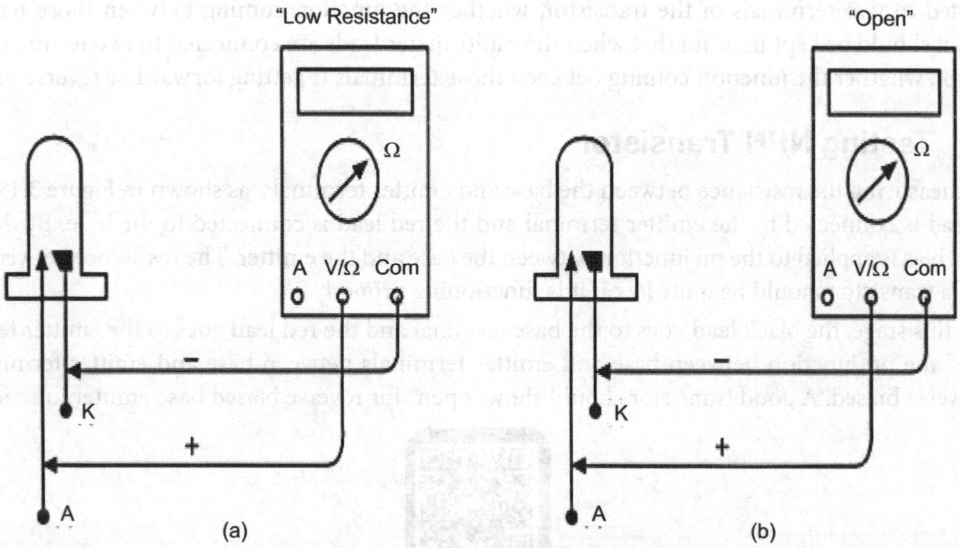

Fig. 3.17: Testing a LED

3.4 BIPOLAR JUNCTION TRANSISTORS

The transistor, also known as a Bipolar Junction Transistor (BJT), is the most common kind of semiconductor and is responsible for many crucial electronic processes. The chances of this device getting faulty are considered to be more than the others. Therefore, knowing its testing procedure is very much important .For its testing, it is necessary to know its construction details. A transistor has three layers: one of one kind of extrinsic semiconductor, followed by two layers of the opposite type, pnp semiconductors are characterised by having a layer of p type semiconductor in the middle of two layers of n type semiconductor. Understanding comparable circuits expressed in terms of diodes is necessary for testing purposes. (Fig.3.18). displays both the npn and pnp transistor equivalents.

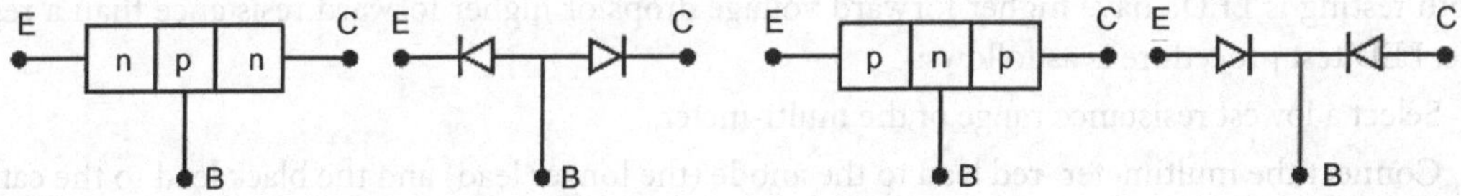

(a) NPN Transistor and its equivalent (b) PNP transistor and its equivalent

Fig. 3.18: Transistor and their Equivalents in Diodes

A NPN transistor can be considered as two diodes with their anodes joined .The joined anode works as the base As illustrated in Figure 1, the base and emitter's p n-junction serves as diode D1, while the base and collector's pn-junction serves as diode D2 (Fig.3.18 (a). In a similar manner, a PNP transistor may be thought of as two diodes with their cathodes joined at base terminal. The Emitter-Base pn-junction acts as diode D_1 while Collector-Base pn -junction acts as diode D_2, as in (Fig. 3.18 (b)). If these equivalent constructions are kept in mind, and the forward reverse bias behaviour to transistors are known, then, it becomes very easy to test a transistor. The resistance between any two terminals can be measured and compared to what it should be in that particular biasing condition, it diode(s) had been there in between those terminals. That is, it should be kept in mind that when the multi-meter leads are connected to two terminals of the transistor, whether the junction coming between those terminals. That is, it should be kept in mind that when the multi-meter leads are connected to two terminals of the transistor, whether the junction coming between those terminals is getting forward or reverse biased.

3.4.1 Testing NPN Transistor

When measuring the resistance between the base and emitter terminals, as shown in Figure 3.19(a), the black lead is connected to the emitter terminal and the red lead is connected to the base. By doing so, forward bias is applied to the pn junction between the base and the emitter. The resistance between these ends of a transistor should be quite low if it is functioning properly.

- In this stage, the black lead goes to the base terminal and the red lead goes to the emitter terminal. So" the pn junction between base and emitter terminals between base and emitter terminal gets reverse biased. A good transistor should show "open" for reverse biased base emitter junction.

Know more about BJT

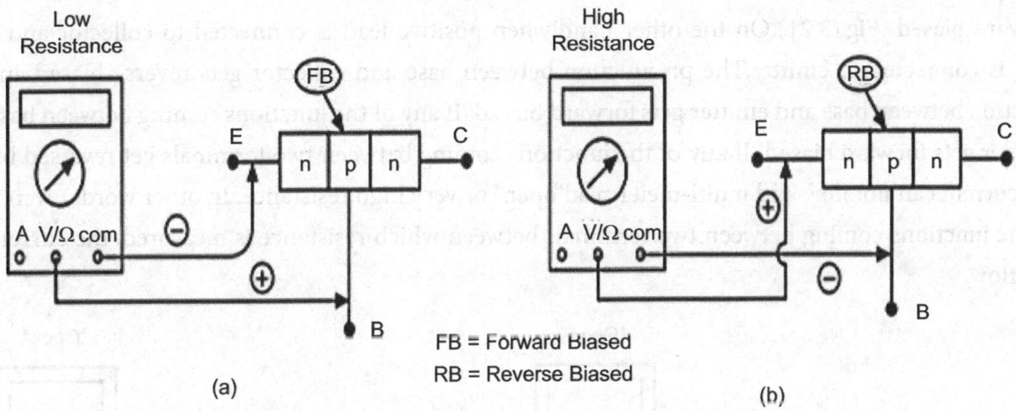

Fig. 3.19: Resistance measurement between Base and Emitter

- Next step have testing of forward and reverse biased resistance between base collector terminals, as (Fig. 3.19) showing connections of red lead of multi-meter to base terminal and black lead to collector terminal. It can be seen from figure that the junction between collector and emitter is forward biased now. A good transistor should show low resistance

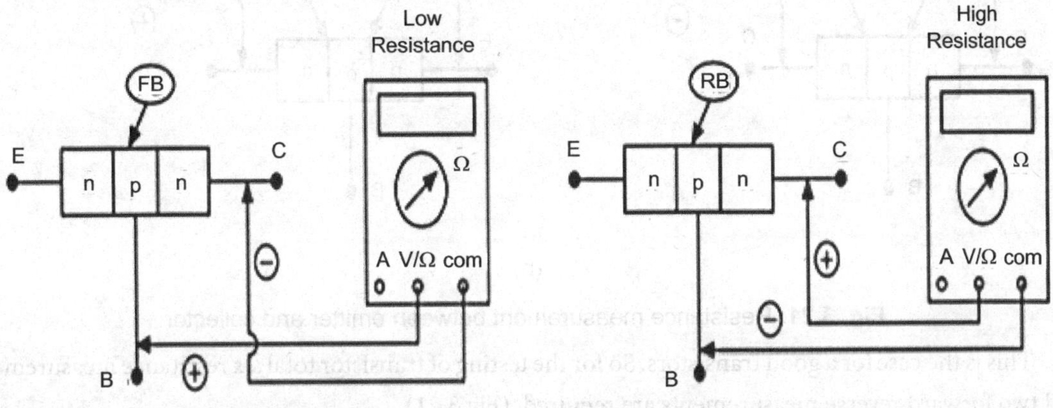

(a) FB = Forward Biased (b) RB = Reverse Biased

Fig. 3.20: Resistance testing between Base and Emitter

Further the red lead of multi-meter to collector terminal while black lead to the base terminal. The pn connection between the base and the collector is now given a reverse bias. Therefore, the multi-meter should show high resistance which ideal value should be infinite . but due to small leakage or thermal current in the reverse bias junction its not possible. If the multi-meter is sensitive enough, this small current is sensed and shown otherwise ,for the good transistor the multi-meter shows "open" circuit. Now measurement of resistance will take place between emitter and collector terminals, in both the directions as in Fig.(3.20).As can be seen from the above figure ,during both the polarities , one of the junction gets reverse biased. When red lead is connected to emitter and black lead to collector, the pn junction between base and emitter gets reverse biased while that between base and collector gets forward biased .Fig.(3.21).On the other hand,when positive lead is connected to collector and black lead is connected to emitter,The pn junction between base and collector gets reverse biased and the junction between base and emitter gets forward biased. If any of the junctions coming between base and emitter gets forward biased. If any of the junctions coming between two terminals get reversed biased. the current can not flow and multi-meter read"open" or very high resistance. In other words ,even if one of the junctions coming between two terminals between which resistance is measured, the current can not flow

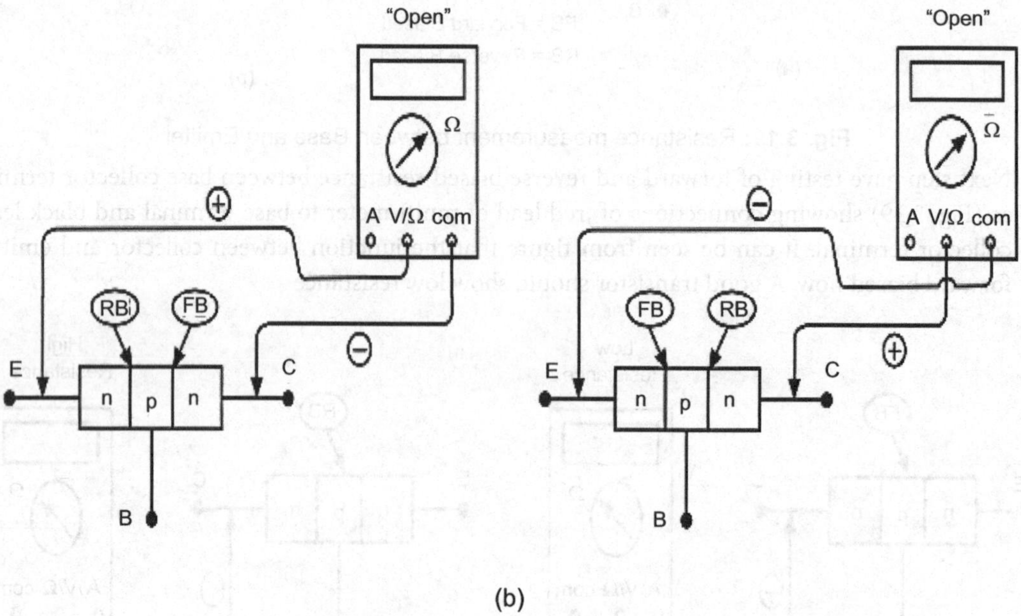

(b)

Fig. 3.21: Resistance measurement between emitter and collector

This is the case for a good transistors. So for the testing of transistor total six resistance measurement and two forward reverse measurements are required .(Fig.3.21)

3.4.2 Testing of PNP Transistor

The testing of a PNP transistor is exactly same to that of NPN transistor. During this test, two resistance measurements are collected between the base and the emitter, two between the base and the collector, and two between the emitter and the collector. The sole distinction may be seen in the connectors between the base and the emitter or collector. As with an NPN transistor, the emitter-collector junction would register as "open" on both probes. For a good transistor, the multimeter will read "open" or extremely high resistance between the base and emitter when the device is reverse biassed, yet it will read low resistance when the leads are switched. And in base-collector terminals in reverse biased ,the multi-meter shows "open" for good transistor or after exchanging the leads connections it shows low resistance.

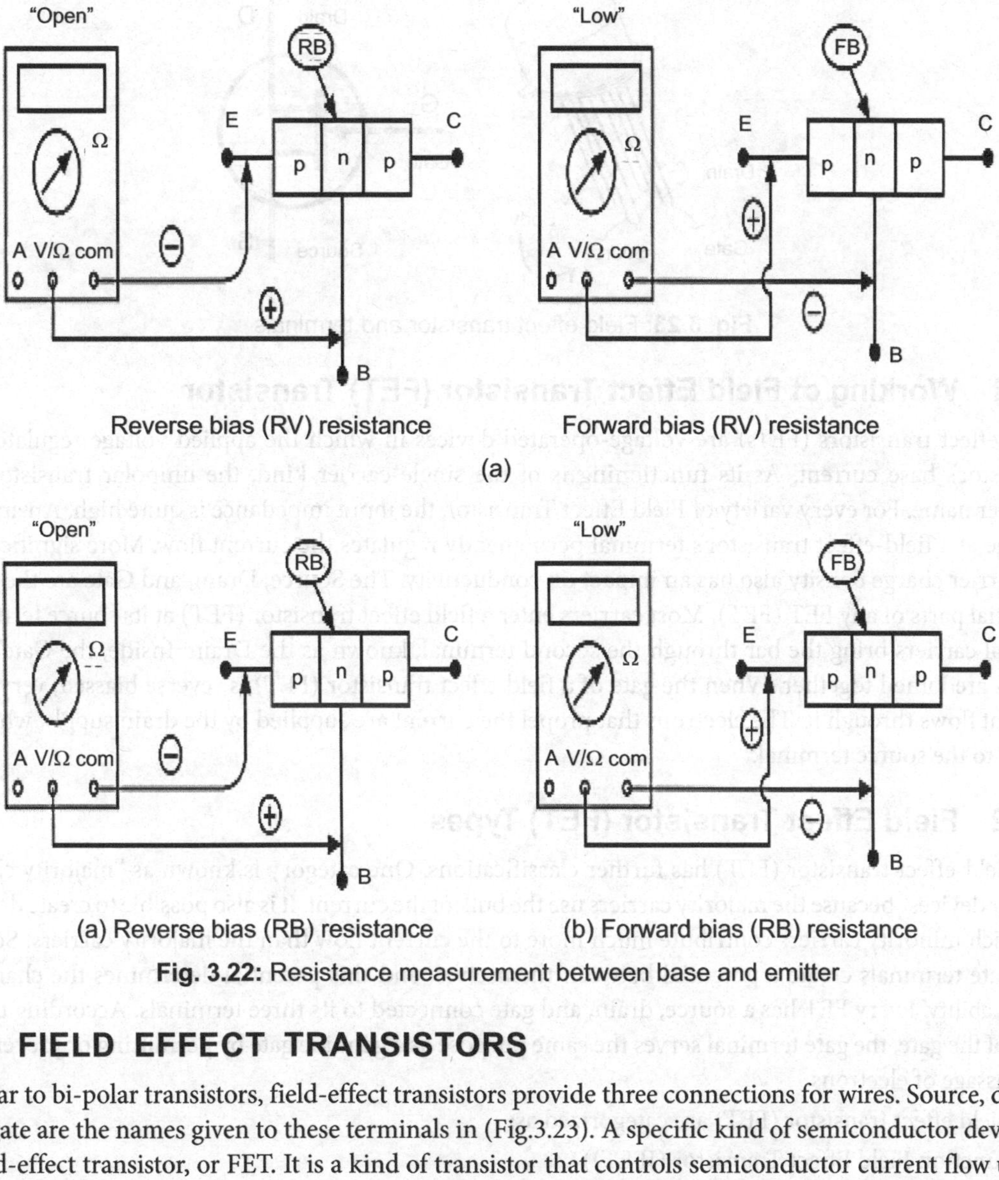

Fig. 3.22: Resistance measurement between base and emitter

3.5 FIELD EFFECT TRANSISTORS

Similar to bi-polar transistors, field-effect transistors provide three connections for wires. Source, drain, and gate are the names given to these terminals in (Fig.3.23). A specific kind of semiconductor device is a field-effect transistor, or FET. It is a kind of transistor that controls semiconductor current flow using an electric field. It is a three-terminal device with a source, gate, and drain. Applying a voltage to the gate of a field effect transistor (FET) changes the conductivity between the drain and source, allowing the current to be controlled.

Know more about JFET

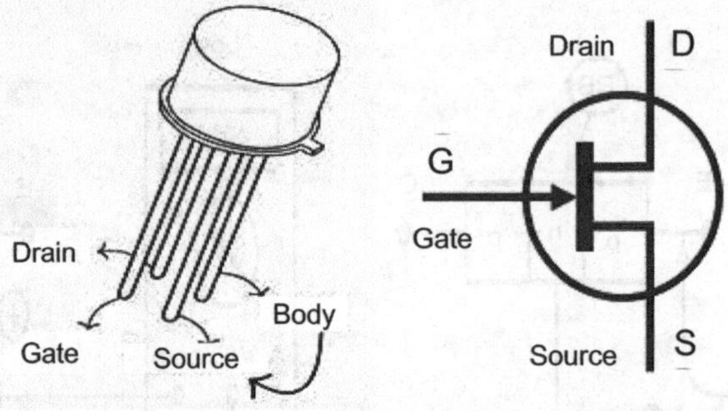

Fig. 3.23: Field effect transistor and terminals

3.5.1 Working of Field Effect Transistor (FET) Transistor

Field effect transistors (FETs) are voltage-operated devices in which the applied voltage regulates the transistor's base current. As its functioning is of the single-carrier kind, the unipolar transistor has another name. For every variety of Field Effect Transistor, the input impedance is quite high. An applied voltage at a field-effect transistor's terminal permanently regulates the current flow. More significantly, the carrier charge density also has an impact on conductivity. The Source, Drain, and Gate are the three essential parts of any FET (FET). Most carriers enter a field effect transistor (FET) at its source terminal. Several carriers bring the bar through the second terminal, known as the Drain. Inside, the Gate's two halves are joined together. When the gate of a field effect transistor (FET) is reverse biassed, very little current flows through it. The electrons that propel the current are supplied by the drain supply, which is wired to the source terminal.

3.5.2 Field Effect Transistor (FET) Types

The field-effect transistor (FET) has further classifications. One category is known as "majority charge carrier devices" because the majority carriers use the bulk of the current. It is also possible to create devices in which minority carriers contribute much more to the current flow than the majority carriers. Source and gate terminals create a potential between themselves, and this potential determines the channel's permeability. Every FET has a source, drain, and gate connected to its three terminals. According to the state of the gate, the gate terminal serves the same purpose as a genuine gate by permitting or preventing the passage of electrons.

Field effect transistor (FET) are categorized as:

- Junction Field Effect Transistor (JFET))
- Metal Oxide Semiconductor Field Effect Transistor (MOSFET)

1. Junction Field Effect Transistor (JFET)

   An electrical switch can be operated by a junction field effect transistor, a type of field-effect transistor (JFET). In this live path, electricity can flow between the source and drain terminals. When a reverse bias voltage is applied to the gate terminal, the channel narrows and the current stops flowing. These JFETs are able to function due to the channels formed between their terminals. It's possible to use either an n-type or a p-type channel. Each type of Junction Field Effect Transistor (JFET) is distinguished by the channel type it uses; an n-channel JFET, for example, uses an n-type channel, and a p-channel JFET, a p-type channel. Field effect transistor (FET) transistors are fabricated using the same N-P-N and P-N-P construction techniques used for bipolar junction transistors (BJT). It is possible to use either an n-type or p-type channel with these JFETs.

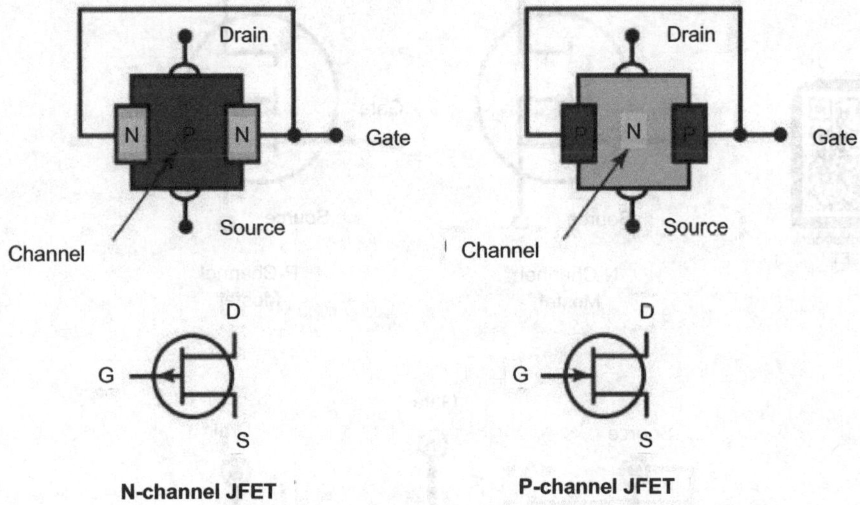

Fig. 3.24: Junction field-effect transistors

It can be either an n-channel JFET or a p-channel JFET, depending on the channel.

- The device is linked to the positive side of an n-channel JFET through its source terminal.
- When compared the drain terminal to the gate terminal of this n-channel JFET, the drain terminal has a larger potential.
- The drain and gate terminals make a connection that is reverse bias. This causes the drain to have a larger depletion zone than the source.
- The majority of charge carriers, electrons, go from the terminal drain to the source in a single direction.
- When the potential at the drain increases, current increases along with the flow of carriers.
- However, by slightly raising the voltages at the source and drain, current flow may be stopped.
- Through the JFET, input voltages may be utilised to control current. This transistor's maximum value is input impedance. There should be no discernible current at the gate terminal when the JFET is operating at peak efficiency.
- In this manner, an n-channel JFET functions. The FET enters p-channel JFET state only when the supply voltage is reversed.

2. Metal Oxide Semiconductor Field Effect Transistor (MOSFET)

   Field Effect Semiconductor Made of Metal Oxide To activate a transistor, a voltage is applied to a channel. Impact of Field on Metal Oxide Semiconductors On the basis of their ways of functioning, transistors may be divided into two broad categories.:

   - Depletion
   - Enhancement

Know more about MOSFET

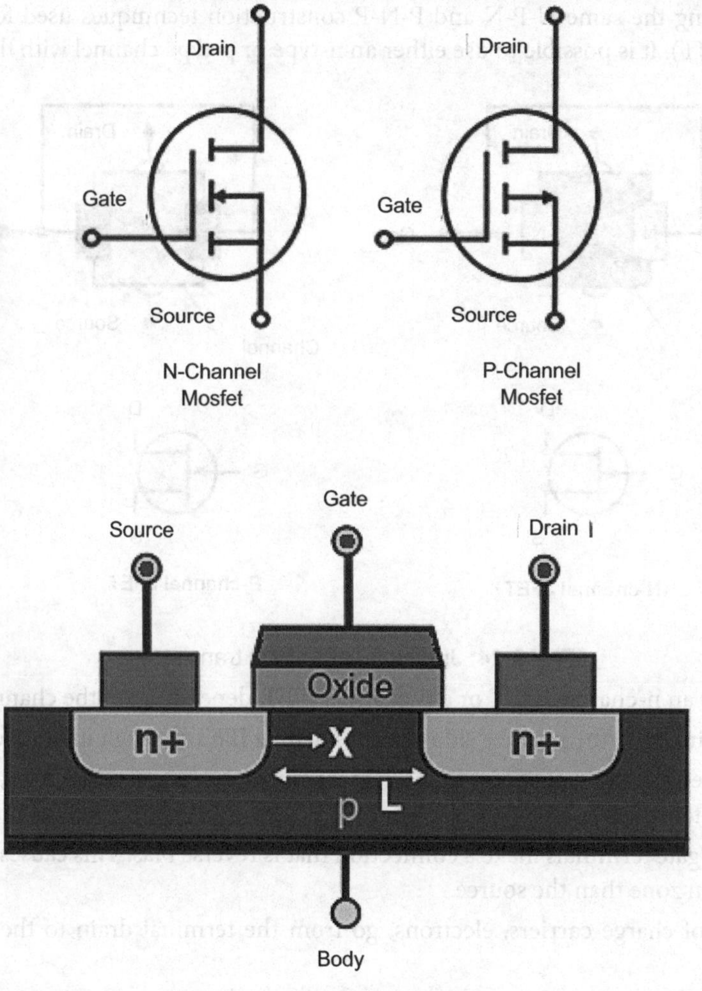

Fig. 3.25: Metal Oxide Semiconductor Field effect transistor (MOSFET)

- In the enhancement mode of a Metal Oxide Semiconductor Field Effect Transistor, the channel is initiated by the gate voltage, but in the depletion mode, the channel already present in the device makes it possible for it to function.
- Both n-type and p-type depletion models are possible for Metal Oxide Semiconductor Field Effect Transistors. The substrate's deposition is the only area of difference. The buildup of the most common carriers leads to the formation of the depletion zone. The conductivity is directly impacted by the extent of the depletion. When the voltage at the gate terminal exceeds a threshold, a channel is created in enhancement mode. The substrate may be n-type if it were P-type, and it could be p-type if it were N-type. An enhancement mode device's type—N-type or P-type Metal Oxide Semiconductor Field Effect Transistor—is determined by channel creation. Modern technology uses enhancement-type Metal Oxide Semiconductor Field Effect Transistors more so than their depletion-type cousins.

3.5.2 Difference Between FET and MOSFET

JFET and MOSFET, the two basic kinds of FET transistors, vary primarily in the following ways.

- Unlike the four-terminal Metal-Oxide Semiconductor Field-Effect Transistor (MOSFET), the three-terminal Junction Field-Effect Transistor (JFET) is also a semiconductor device (Metal Oxide Semiconductor Field-Effect Transistor).
- JFETs can only function in a depletion mode. A MOSFET, on the other hand, may function both in the enhancement and depletion modes.
- Higher input impedance makes MOSFETs more resistant. When comparing prices, MOSFET is clearly more costly than JFET.
- Due to its high input impedance, FET transistors are often utilised in and as input amplifiers in electronic voltmeters, oscilloscopes, and other measuring apparatus. They take up very little space, which improves their usefulness in conjunction with other gadgets.

3.5.3 Testing of Field Effect Transistor

Field-effect transistors need a little more careful and extensive testing approach than bi-polar transistors. The following must be determined before putting the FET through its paces:

- The purpose of the device is the same whether it is a junction field-effect transistor (JFET) or a metal-oxide semiconductor field-effect transistor (MOSFET).
- What is the type of channel 'n-channel' or' p-channel '
- In case of (MOSFET), is it may be enhancement type or a depletion type

Comparison is made between a junction gate and a regular insulated gate. Simply put, a field effect transistor (FET) is a "ON" device similar to a vacuum tube. Both of these gadgets require a bias-off adjustment. Newer, more advanced, insulated gate Conversely, a field effect transistor (FET) is an inherently "OFF" device that requires a bias. By converging at a junction, an ohmmeter can be used to test a field-effect transistor (FET).(Fig.3.26) shows connectors.junction field effect transistor resistance measurements (JFET). Measurements in both directions are taken only between the gate and the source or the gate and the drain. It doesn't matter which way you stick your ohmmeter probes into the source and drain connections, you'll always get the same reading. Remember that the resistance values will vary with the range because of the non-linear nature of the gate source and gate drain junctions.

Using an ammeter, a junction field effect transistor's functionality can be quickly verified. An n-channel device is depicted in (Fig.3.27(a)), while a p-channel device is depicted in (Fig.3.27(b)). The metre will show a few milliamperes.

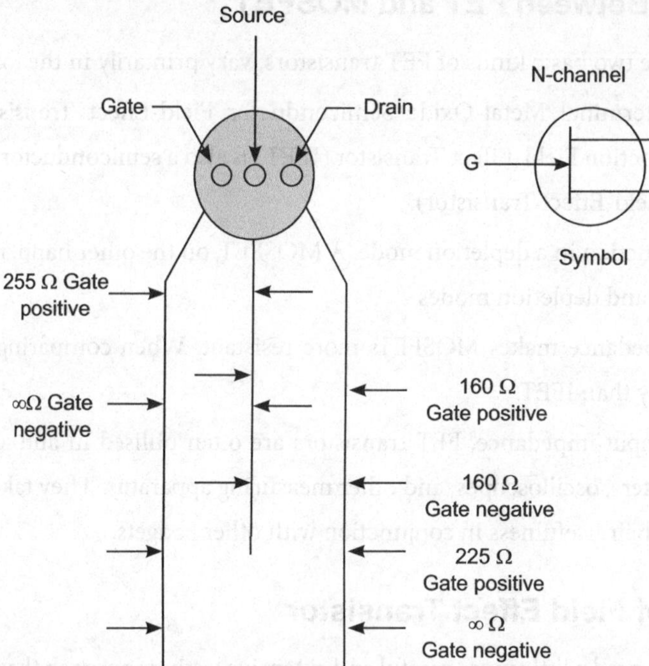

Fig. 3.26: Junction resistance readings of a junction field effect transistor (JFET)

if the gate is first connected directly to the source (rather than as indicated). The drain circuit includes a 1K resistor that limits the flow of current to a safe level. In this case, the gate electrode is connected to a 10 M resistor, creating a reverse bias with respect to the channel junction. As a consequence, the drain current in the circuits presented approaches zero for the majority of devices and the channel width shrinks.

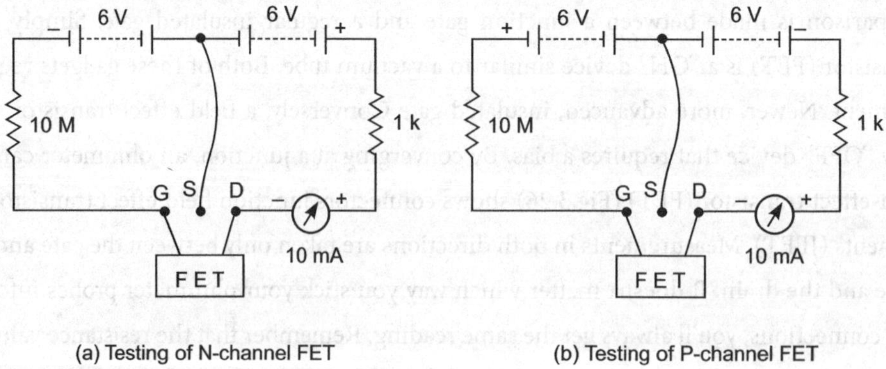

Fig. 3.27: Testing method for Junction Field effect transistor (JFET)

To apply voltage to the gate circuit, a high value resistor must be used due to the circuit's high resistance (10 M). If you want to test a gadget but have no idea how it is wired, all you have to do is discover two points where a modest current will flow in either direction. These are the water supply and waste lines. The gate electrode (third electrode) must be a one-way current source, with current flowing solely from one electrode to the other. If conduction happens when the gate is positive, one has an n-channel device; if conduction occurs when the gate is negative, one has a p-channel device. The drain and source electrodes are not readily distinguishable, although they may be used interchangeably for electrical purposes. One thing to keep an eye on is the amount of moisture that accumulates in the soldering fluxes from exterior leakage routes. A Field Effect Transistor (FET) with such pathways will falsely be suspected as being broken. Therefore, an external leakage channel (not inside the FET) may develop. Therefore, before swapping out leaky Field effect transistors (FETs), the PC board must be meticulously cleaned (FET). Typewriters have very efficient erasers. When soldering or desoldering a junction FET or an insulated-gate protected metal-oxide-semiconductor field-effect transistor (FET), no special care is needed as long as the junction temperature is kept below 125 degrees Celsius. The insulation acts as a tiny capacitor between the gate and the channel in insulated gate Field effect transistors (FETs), which is something to keep in mind. Due to their fragility, the insulation and gate are easily harmed (Fig.3.28). Keep in mind that the static charge from the fluorescent lights could cause serious injury if the gate lead were left unlocked. This could happen if, prior to installation, the Field Effect Transistor (FET) is taken out of its protective casing and set on a table with its shorting wire cut. In many cases, a shorting wire is nothing more than a short length of wire that is wrapped around each individual lead. When handling MOSFETs, it's important to follow these precautions.

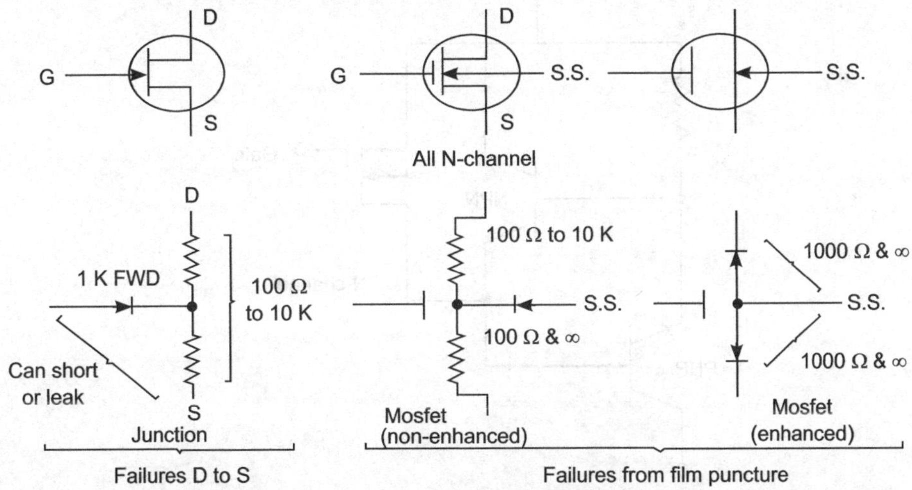

Fig. 3.28: Failures in field-effect transistors

(a) The MOSFET device leads must remain shorted until they are ready to be connected to the circuit.
(b) It is essential that the user's body and hand be at ground potential before beginning.
(c) The tips of soldering guns need to be grounded during the soldering process.
(d) Never add or remove a MOSFET from its circuit while it is powered on.

Testing the MOSFET with an ohmmeter is possible. We use the ohmmeter's maximum resistance setting to test the gate-to-source insulation. Each polarity of the test voltage often yields the same almost limitless value. It is common to find a very low drain-to-source resistance, and ohmmeter readings that are unaffected by the direction of the test voltage. Keep in mind that the most typical MOSFET failure scenario is a short circuit between the gate and source or drain and source. Put another way, there is no such thing as a self contained part.

3.5.4 Insulated Gate Bipolar Transistor (IGBT)

Prior to the invention of IGBTs, power MOSField effect transistors (FETs) were used for medium-to-low voltage applications requiring fast switching, while bi-polar power transistors and thyristors were used for medium-to-high voltage applications requiring high current conduction. Power MOSFETs have excellent fast-switching performance and a simple gate control circuit design. However, when the breakdown voltage is higher than 200 V, the on-resistance rises nearly as quickly, which is a major drawback. The bipolar power transistor's low forward voltage drop gives it excellent on-state characteristics, but its complex base control circuit and slow switching operation make it less desirable than a field-effect transistor made from metal-oxide semiconductor (MOSFET). The IGBT combines the benefits of both the transistor and the diode.

In(Fig.(3.29)), we see the IGBT structure, which is a combination of the power MOSFET and the bi-polar power transistor. MOS gate at the input, wide base PNP transistor at the output. Once the PNP transistor is turned on, the base driving current will travel through the input channel. In contrast to the PNP transistor, the NPN transistor can be disabled by forming a short circuit between its base and emitter and the metal of the MOSFET's source. Latching can occur due to the PNPN thyristor's four layers, which include the PNP transistor and the NPN transistor. Because it lacks a parasitic reverse diode as a power MOSFET does, a quick recovery diode must be linked to it when required.

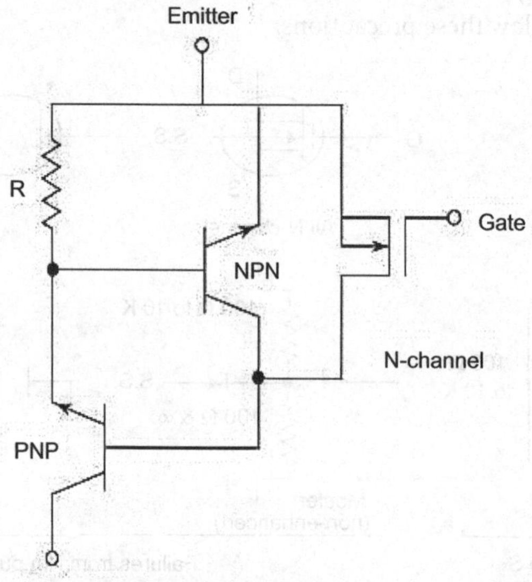

Fig. 3.29: Structure of IGBT (Insulated Gate Bipolar Transistor)

3.6 THYRISTORS

Thyristor is semiconductor device and having three types

- Rectifier controlled by silicon
- Triac
- Connector thyristor with four leads

Know more about Thyristors

Power control circuits rely heavily on thyristors. They excel in controlling alternating current and are hence ideal for use in dimmer switches, speed regulators, temperature controllers, and invertors. They are also used in DC power supply for over-voltage protection.

The thyristor may be conceptualized as a two-transistor combination structure, and it is really a four-layer PNPN device (Fig. 3.30). Both the NPN and PNP transistors are linked in series across each other. Both the NPN collector and PNP base are connected to each other via a coupling network. However, a trigger pulse applied to the thyristor's gate can switch it into the low-resistance forward conducting state from its inactive off state. If the current through the thyristor is reduced to below its holding current value or if it is reverse-biased, the thyristor will continue conducting after being triggered into conduction. Consequently, the voltage-current characteristics of the thyristor are very nonlinear (Fig. 3.31).The triac functions similarly to two thyristors paralleled with a common gate. This indicates that the gadget may function in either a conducting or insulating capacity. Both positive and negative gate signals may initiate its conduction, hence its directionality is immaterial.

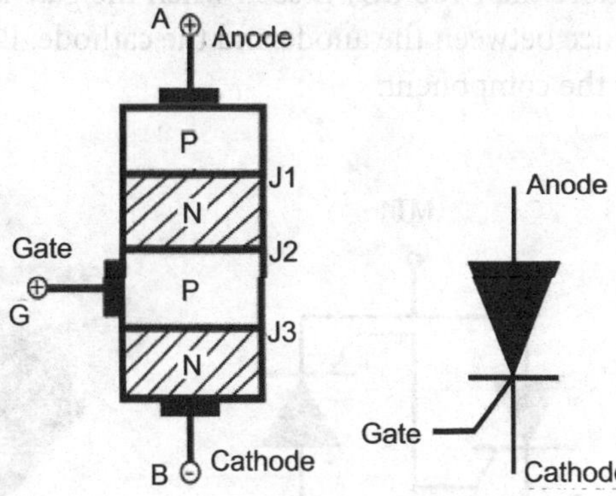

Fig. 3.30: Silicon-controlled rectifier

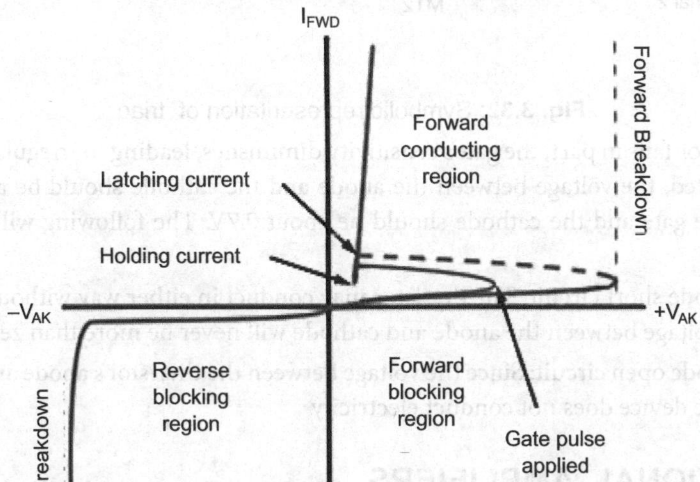

Fig. 3.31: V-I characteristics of SCR

The power output of a motor speed controller or light dimmer is controlled by timing the thyristor's trigger pulses in relation to the AC power's zero crossings. When the thyristor is turned on at the beginning of the cycle, a lot of energy is transferred to the load. Similar to how a small amount of power is sent to the load if the thyristor is activated late in the cycle. Since thyristors are either "on" or "off," there is almost no power loss when compared to conventional variable resistors. Thyristors and triacs may fail if exposed to extreme heat or rapid temperature changes. These causes of failure often result in open circuit or short circuit between terminals, resulting in the total breakdown of the device.

These malfunctions may be examined using an ohmmeter. The diode's gate-cathode is one such example. Indicating a low resistance when the gate is positive in relation to the cathode On the other hand, a significant resistance (more than 100 kQ) is seen when the gate is negative with respect to the cathode. There is a lot of resistance between the anode and the cathode. Before taking readings with an ohmmeter, you should desolder the component.

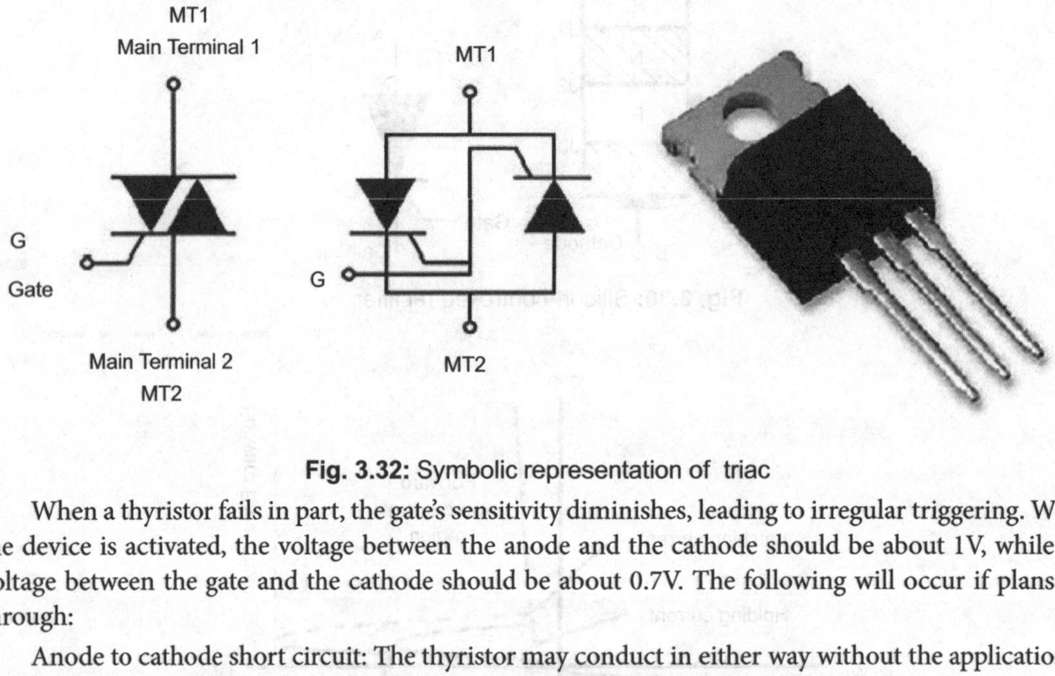

Fig. 3.32: Symbolic representation of triac

When a thyristor fails in part, the gate's sensitivity diminishes, leading to irregular triggering. When the device is activated, the voltage between the anode and the cathode should be about 1V, while the voltage between the gate and the cathode should be about 0.7V. The following will occur if plans fall through:

Anode to cathode short circuit: The thyristor may conduct in either way without the application of a gate signal. The voltage between the anode and cathode will never be more than zero.

Anode to cathode open circuit: Since the voltage between the thyristor's anode and cathode remains high at all times, the device does not conduct electricity.

3.7 OPERATIONAL AMPLIFIERS

A differential voltage input can be transformed into a single-ended voltage output using an analogue circuit component called an operational amplifier (op amp). Normal operational amplifiers have one low-impedance output and two high-impedance inputs. An inverting input is represented by a minus sign (-), while a non-inverting input is represented by a plus sign (+). Operational amplifiers have many analogue applications in signal processing, power distribution, and control in addition to their primary purpose of differential voltage amplification.

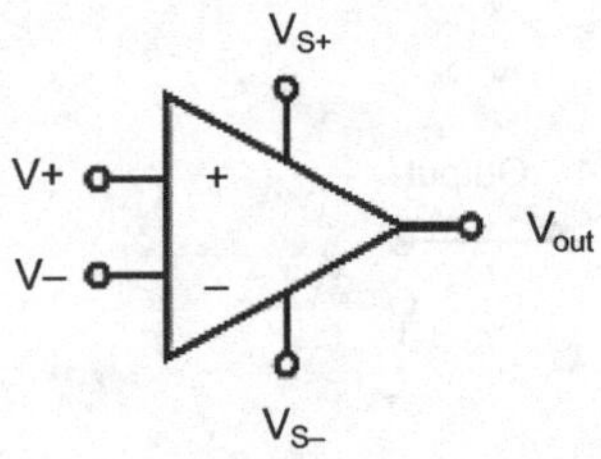

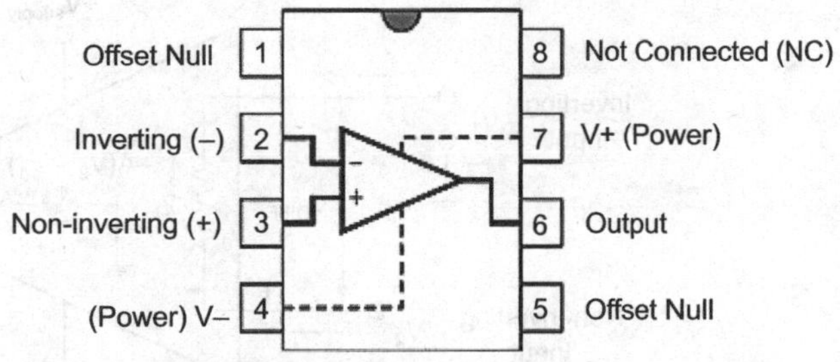

Fig.3.33: Symbol and Integrated circuit of operational amplifier

3.7.1 Operational Amplifier Classification

Operational amplifier classified in four categories:

1. Voltage amplifier
2. Current amplifier
3. Trans conductor amplifier
4. Trans resistance amplifier

An operational amplifier (or "Op-Amp" for short) is a special kind of integrated circuit that can perform a wide variety of linear, non-linear, and even mathematical operations. An operational amplifier is a direct-coupled, high-gain amplification device. The use of op-amps is not limited to either AC or DC current (DC).

3.7.2 Construction of Operational Amplifier

An operational amplifier consists of output stage, level translator, and differential amplifiers. Th op-input amp stage is a differential amplifier, so it requires two input terminals. The phrase "inverting terminal" refers to the one that switches polarity, whereas "non-inverting terminal" describes the other. The phase connection between the inputs and the outputs is taken into account while naming the terminals.

3.7.3 Characteristics of Operational Amplifier

The following are some of the most crucial aspects of an operational amplifier:–

- Output offset voltage

When the differential input voltage is 0, the voltage at the output of an op-amp is called the output offset voltage. In the case where the potential difference between the two input terminals is positive, the resulting dc voltage from the offset output is also positive, and vice versa if the difference is negative.

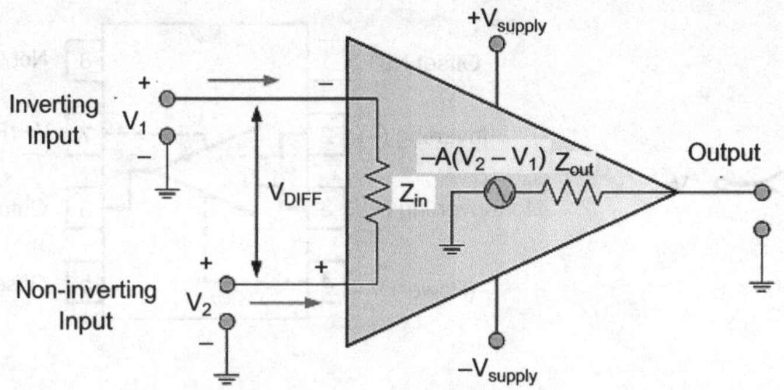

Fig. 3.34: Output offset voltage in op amp

- Common Mode Rejection Ratio(CMRR)
  Divide the differential gain (A_D) by the common mode gain to compute it (A_{CM}). The CMRR for the 741C is around 90 dB approximately

$$CMMR = \frac{A_D}{A_{CM}}$$

- Slew Rate
  The quickest rate at which an operational amplifier reacts to a step in the input voltage is known as the output voltage slew rate.

$$\text{Slew rate (SR)} = \text{Maximum of } \frac{dv_0}{dt}$$

Where, V_0 is the output voltage. In general, slew rate is measured in either V/µ Sec or V/m Sec.

3.7.4 Types of Operational Amplifiers

The op-amp symbol is a triangle with two inputs and a single output.
There are two kinds of operational amplifiers, the "ideal" kind and the "practical" one.
What follows is a comprehensive analysis of these factors.–

Ideal Op-Amp

An perfect op-amp doesn't really exist; it only exists in theory.

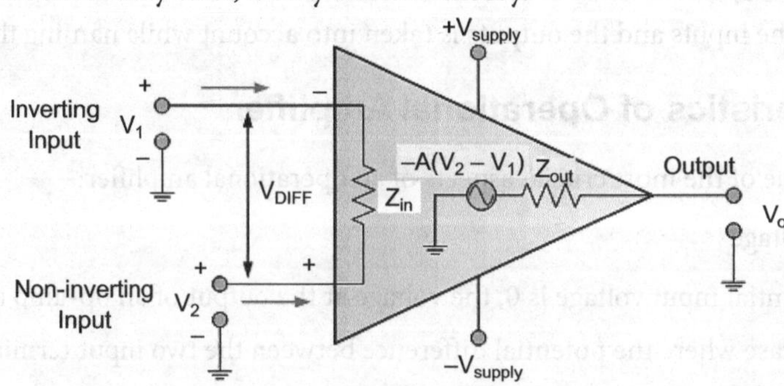

Fig.3.35: Equivalent circuit of an ideal operational amplifier

- An ideal operational amplifier having following characteristics:-
- output impedance $Z_0 = 0\Omega$
- Input impedance $Z_1 = \infty\ \Omega$
- open loop voltage gain AV = ∞V, It means if the differential input voltage is 0V, then the output voltage will be 0V
- Bandwidth is infinity. It means ,an ideal operational amplifier will amplify the signals of any frequency without any attenuation
- Common mode rejection ratio is infinity.
- Slew rate is infinity. It means ,the ideal op-amp will produce a change in the output instantly

Practical Op-Amp

- In practise, op-amps depart from their ideal features and are not flawless. This is due to manufacturing defects. The following diagram depicts the analogous circuit of a real-world op-amp

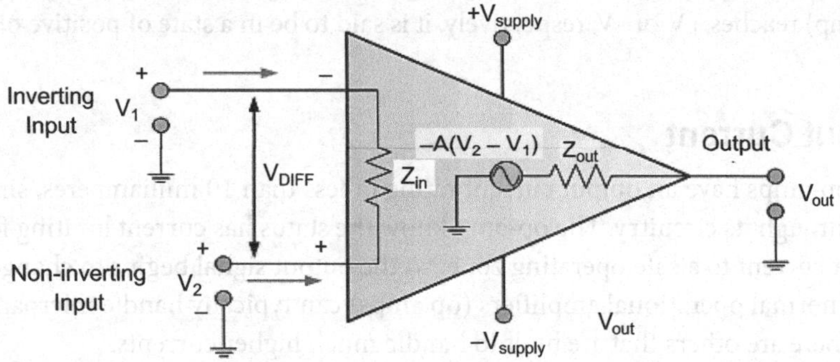

Fig. 3.36: Practical operational amplifier

A practical op-amp exhibits the following characteristics –

- Zi, the input impedance, is on the megaohm level.
- As for the output impedance, Z0, it's on the order of a few ohms.
- Gain in open loop voltage, Av, is large.

The following criteria should be considered while selecting a usable operational amplifier (op-amp):

1. Maximum input impedance (Zi) is desired.
2. Z0, the output impedance, should be kept as low as feasible
3. The open-loop voltage gain, Av, should be maximized.
4. Lowering the output's offset voltage is a top priority.
5. Bandwidth in operation should be maximized.
6. Maintain the highest possible CMRR.
7. Slew rate should be maximized.

3.7.5 Power Supply Requirements for Operational Amplifier

Like every other kind of transistor amplifier, op-amps need a direct current (DC) power source. The power supply needs enough filtering and precise voltage control for the op-amp to operate effectively.

The op-power amp's + V and - V lines should be connected to the positive and negative ends of the supply voltage, respectively. Voltages from the supply, whether positive or negative, are frequently symmetrical, meaning that they have opposite signs but are still the same magnitude.

Op-amps require a supply voltage between +15 V and -15 V. On the other hand, this isn't always the case. To find out what kind of power source the op-amp in question needs, you should consult the manufacturer's data manual for that op-amp. It's important to remember that op-amp power wires aren't always depicted on circuit diagrams. So we can say that the op-amp needs a DC voltage in order to function..

3.6.6 Output Voltage

The op-voltage amp has a restricted swing, similar to that of common transistor amplifiers. The op-positive amplifier and the magnitude of the negative DC power supply voltage set the upper and lower bounds of the output voltage swing. Typically, an output amplifier's (op-amp) voltage input and output range is limited to the supply rails' positive and negative voltages. When the output of an operational amplifier (op-amp) reaches +V or -V, respectively, it is said to be in a state of positive or negative voltage saturation.

3.7.7 Output Current

Most common op-amps have an output current rating of less than 10 milliamperes, since this is all that can safely flow through its circuitry. The op-amp know the status has current limiting features that limit their own output current to a safe operating zone, so the output signal begins to change if more current is drawn. While normal operational amplifiers (op amps) can typically handle microampere currents at the output pin, there are others that are built to handle much higher currents.

3.7.8 Typical Op-Amp Circuits

Before outlining the process for troubleshooting op-amps, it's helpful to consider the various circuit configurations in which the op-amp is used to perform a variety of purposes, such as a precision amplifier, summing device, integrator, distinguisher, generator, active filter, comparator, etc.

The performance of an operational amplifier in a negative feedback circuit is significantly influenced by the size of the external components. Here are some common setups for operational amplifiers with various types of feedback:

Inverting amplifier

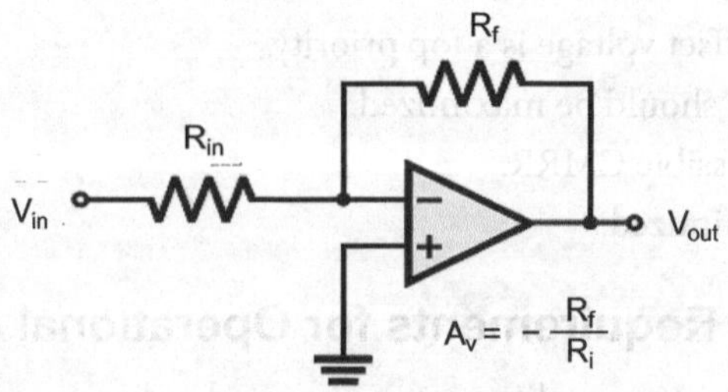

$$A_v = -\frac{R_f}{R_i}$$

Fig. 3.37: Inverting amplifier

$$\text{Gain } A = \frac{V_{OUT}}{V_{IN}} = -\frac{R_f}{R_i}$$

where A is the closed loop gain of the amplifier.

A negative sign indicates that the output signal is 180 degrees out of phase with the input signal, and the only other factors that affect inversion are the values of Rf and Ri in the feedback pathway. The closed-loop gain of an ideal amplifier is decreased when non-resistors are used in the feedback circuit.

Non-inverting Amplifier

In the case of a non-inverting amplifier, the closed-loop gain is defined as

$$A = 1 + \frac{R_2}{R_1}$$

If R_2 is 20k and $R_1 = 2$ k, then $A = 1 + 10 = 11$

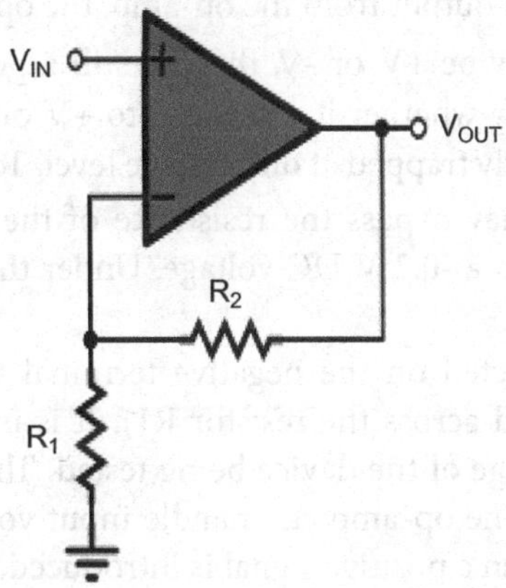

Know more about OP-AMP

Fig. 3.38: Non-inverting amplifier

3.8 FAULTS DIAGNOSIS IN OP-AMP CIRCUITS

Operational amplifiers in modern integrated circuits are very reliable. However, many various types of internal errors, such as shorts and open circuits, may still arise inside the IC itself. Therefore, during fault diagnostics, it is crucial to quickly pinpoint the issue to either the IC or another component. Soldering should only be done after any external problems, such as open circuits or shorted pins or tracks, have been ruled out .Fault diagnosis techniques for op-amp circuits are grounded in the op-amp's aforementioned characteristics as well as the author's prior practical experience. Troubleshooting procedures for operational amplifier are as follows

(a) Check the (+) and (-) pins to see what the voltage is like in inverted and non-inverted states. A difference of more than 10 mV is unacceptable. If the voltage is not within this range, investigate the surrounding electronics. But it isn't to say the op-amp isn't broken in some other way. It's possible that the op-amp is faulty, but it's too soon in the problem diagnostic procedure to say for sure.

(b) Determine the voltage at the operational amplifier's (+) and (-) power inputs.

The op-amp package itself, not simply the PC (Printed Circuit) board, should be measured. Check for the following issues if the supply is available on the PC board but not the IC pins:

1. Perhaps the PC board track is broken or unreliable.
2. The PCB's feed-through hole isn't fully metallized the way through.
3. Perhaps the socket is broken if the gadget is already inserted.

To find out why there is no voltage on the PC board, you must first trace the power supply voltage on the board.

(c) There may be an issue with the output circuit if the op-power amp's supply is sufficient. As was mentioned before, the op-amps' output current drops dramatically when a load is connected to the amplifier's terminals (only 1-10 mA).This hypothesis should be ruled out by isolating the output pin from the rest of the PC board and looking out for the following signs.

(d) To check the voltage at the output pin, we must first take out everything but the feedback circuit. The voltage may either be positive (+V) or negative (-V), since the output will fluctuate in wattage to the extent that the power source permits (-V) (-V). When the output voltage is still 0.0 volts after isolating the digital input and output from the op-amp, the op-amp is defective.

Even if the output voltage may be +V or -V, there is still a potential that the device is damaged. Testing the op-range amp's to verify whether it can travel to +V or -V is critical since, in the case of a failure, its output can be permanently trapped at one voltage level. To make this happen, take these steps.

(i) With a shorting lead, you may bypass the resistance of the resistor Ri. An op-inverting amp's terminal will be connected to a -0.2 V DC voltage. Under these circumstances, the output must positively saturation.

(ii) A positive DC signal is injected on the negative terminal to force the output to the -V level (disconnect the shorting lead across the resistor R1). It is important that the signal strength is within the safe operating range of the device being tested. The manufacturer's device data sheets should include this details. The op-amp can handle input voltages of several hundred millivolts without being damaged. When a positive signal is introduced, the output will eventually settle at a negative value.

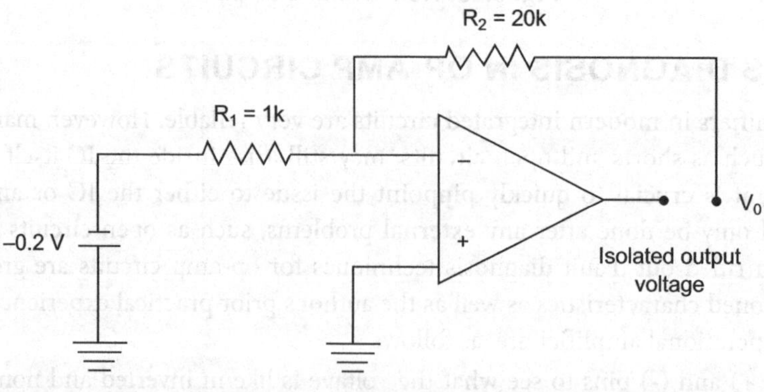

Fig. 3.39: Op-amp circuit with isolated output of the op-amp

A malfunction exists in the device if the output does not saturate at a positive voltage or a negative voltage, respectively. If the op-output amp's can flip between positive and negative voltages, it is probably functional.

(d) The feedback resistor R must be connected even when the output circuit is isolated in order to verify that the device is operating as intended (Fig. 3.40). This will relieve the op-amp of any load on the output circuit. The op-amp is functioning if and only if the measured output voltage is greater than 4.0 volts. If the op-voltage amp's is still below +4.0 volts, it's broken and needs to be replaced.

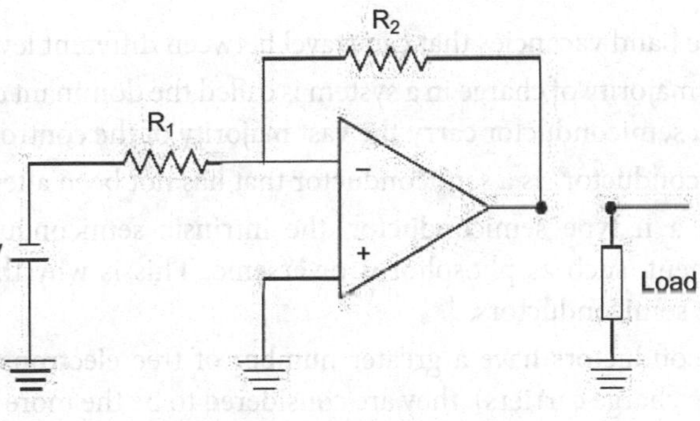

Fig. 3.40: Isolated load circuit from the op-amp circuit

(e) In the absence of the output circuit, an op-amp circuit should continue to function normally; if it does, the issue lies with the load circuit. Verify that the op-amp output circuit on the PCB and any other connected circuits are not shorted to ground.

Most op-amp circuits you encounter should be easily diagnose and analyse using these methods.Even while not every op-amp circuit is an inverting amplifier, the procedures outlined here might help you figure out whether the op-amp in question is broken. When there are several operational amplifiers (op-amps) in a circuit, knowing this information is crucial.

Know more about component testing

UNIT SUMMARY

- Materials used in semiconductor devices have a conductivity that falls between between that of a conductor and an insulator.
- Semiconductors provide improvements in communication, computing, healthcare, military systems, transportation, renewable energy, and a wide range of other applications.
- An N-type semiconductor is used when there are a lot of free electrons since this increases conductivity. A P-type semiconductor is employed when the inductance is higher yet there are fewer free electrons.
- Due to their compact nature, semiconductor devices cannot dissipate heat as quickly as it is generated within them, raising the internal temperature over time might lead to permanent damage.
- The Bipolar System Charge-type carriers flowing across forward- and reverse-biased junctions are essential to the devices' operation.
- Bipolar devices that are often utilised include transistors, diodes, mono junction transistors, thyristors, logic integrated circuits (ICs) like TTL, and linear ICs.
- In unipolar devices, the electrostatic field between the gate and the source, or the gate and the substrate, determines the direction of the current flow.

- Charge carriers, such as free electrons, that transport negative charge from one location to another are called "negative charge carriers."
- The electrons in a free electron are those that are not bound to the parent atom and may thus travel freely.
- Hole-like positive charge carriers are the ones that transport positive charge from one location to another.
- "Holes" are valence band vacancies that can travel between different levels of the band.
- One that holds the majority of charge in a system is called the dominant charge carrier. The majority charge carriers in a semiconductor carry the vast majority of the control or current.
- An "intrinsic semiconductor" is a semiconductor that has not been altered from its natural state
- In order to make a n type semiconductor, the intrinsic semiconductor must be doped with a pentavalent element, such as phosphorus or arsenic. This is why there are so many unpaired electrons in n-type semiconductors.
- Since n-type semiconductors have a greater number of free electrons (negative charge carriers) than holes (positive charge carriers), they are considered to be the more "conductive" type (positive charge carriers).
- The number of holes in a p-type semiconductor is greater than the number of free electrons in that material.
- By combining the intrinsic semiconductor with trivalent elements, such Boron or Gallium, a p type semiconductor is created. An abundance of holes characterizes p-type semiconductors.
- The assembly process may have been botched, there was too much voltage, current, or power, or there was electrical interference. All of these factors could cause a semiconductor device to malfunction.
- Diodes are essential component of electronic devices in which electric current flows in only one direction and have metal casings at both the cathode and anode ends.
- Using an ohmmeter to measure a diode's forward and reverse resistance allows for a fast and simple test.
- The resistance of a typical signal diode or rectifier is low in the forward direction and high in the reverse direction.
- If you want accurate results from your ohmmeter, you must always disconnect one end of the diode from the circuit before measuring the other end.
- Above a specific applied voltage, the zener diode begins conducting strongly in the opposite direction.
- Varactor diodes are silicon diodes that function across a broad range of reverse voltages as a variable capacitor. A varactor's nominal capacitance may range from 1 pF to 500 pF
- A varistor is a kind of semiconductor whose resistance changes depending on the voltage supplied..
- The efficiency of a photo-diode may be determined by recording the voltage rise across a resistance.
- While the screen does not indicate how long it takes for the device to switch, it does validate that it can handle the specified amount of current and should function appropriately in the circuit.
- A tunnel diode is a pn junction with a negative resistance gap.
- Checking the current at which the tunnel diode switches on and off is a useful first step in diagnosing many problems.

- One of the most essential components in electrical circuits, diodes serve a number of functions, including switching, rectification, and protection.
- Variations in forward voltage drop between 0.2 V and 0.4 V are typical for Germanium diodes. This is because the 'Germanium' substance has a relatively low break over voltage.
- Zener diode are designed to have very sharp reverse breakdown voltage. The rating of a zener diode is actually this reverse voltage at which the breakdown occurs.
- Testing a zener diode with a multi-meter is just like measuring a normal diode. It involves measuring its forward and reverse bias resistance
- A forward bias should register as a voltage drop of 0.6 V to 0.8 V on a multimeter. Under a reverse bias exceeding its breakdown voltage.
- If the multi-meter shows the applied Dc voltage as the measurement,then,it means that the diode is open. If the multi-meter shows zero voltage,that means diode is short
- In BJT transistor consists a type of extrinsic semiconductor layer sandwiched between two layers the opposite type of extrinsic semiconductor .
- When a semiconductor with a P type layer lies between two semiconductor layers with a n type layer, the resulting device is a "pnp"transistor. It is possible to think of an NPN transistor as two diodes with their anodes linked.
- Similar to the bi-polar transistor, the field-effect transistor has three terminals for external connections. They are formally known as the "source," "drain," and the "gate," respectively. The Field effect transistor (FET) utilises an electric field to control current flow within a semiconductor.
- Applying a voltage to the terminal of a field-effect transistor is a constant approach to control the material's conductive properties
- For straight forward gate control circuit design and top-notch performance, a power MOSFET is a great choice.
- Power control circuits make considerable use of thyristors. They function admirably in AC power-regulating gadgets like light dimmers, motor speed controllers, thermostats, and inverters.
- An operational amplifier consists of a level translator, an output stage, and one or more differential amplifiers. An op-amp has two terminals because it features a differential amplifier as its input stage. The terms "inverting terminal" and "non-inverting terminal," respectively, set them apart.

EXERCISES

Multiple Choice Question

1. A semiconductor is formed by .... bonds.
   (a) Covalent (b) Electrovalent
   (c) Co-ordinate (d) None of the above
2. The temperature coefficient of resistance for a semiconductor is..
   (a) Positive (b) Zero
   (c) Negative (d) None of the above
3. The ubiquitous semi-conductor that everyone uses is...
   (a) Germanium (b) Silicon
   (c) Carbon (d) Sulphur

4. A semiconductor has generally .... valence electrons.
   (a) 2 (b) 3
   (c) 6 (d) 4
5. Under typical conditions, pure germanium has a resistivity of........
   (a) 6 × 104 (b) Ω cm
   (c) 60 (d) Ω cm
6. The addition of a pentavalent impurity to a pure semiconductor results in the formation of.........
   (a) An insulator (b) An intrinsic semiconductor
   (c) p-Type Semiconductor (d) n-Type Semiconductor
7. A JFET has three terminals, namely
   (a) Cathode, Anode, Grid (b) Emitter, Base, Collector
   (c) Source, Gate, Drain (d) None Of The Above
8. In many ways, the operation of a JFET is like that of a.............. valve.
   (a) Diode (b) Pentode
   (c) Triode (d) Tetrode
9. The gate of a JFET is .... biased
   (a) Reverse (b) Forward
   (c) Reverse As Well As Forward (d) None Of The Above
10. The width of the conducting channel grows in response to an increase in the reverse bias on the gate of a JFET.
   (a) Is Decreased (b) Is Increased
   (c) Remains The Same (d) None Of The Above
11. MOSFET can be operated with ....
   (a) Negative Gate Voltage Only (b) Positive Gate Voltage Only
   (c) Positive As Well As Negative Gate Voltage (d) None Of The Above
12. The pnp transistor›s common base configuration is similar to the............ of the JFET.
   (a) Common Source Configuration (b) Common Drain Configuration
   (c) Common Gate Configuration (d) None Of The Above
13. Find the device with the lowest input impedance.
   (a) JFET (b) MOSFET
   (c) Crystal diode (d) Ordinary transistor
14. The pinch-off voltage of a JFET is about ....
   (a) 5 V (b) 0.6 V
   (c) 15 V (d) 25 V
15. It is important to note that a FET›s pn junctions are located on its sides.
   (a) Three (b) Four
   (c) Five (d) Two
16. A JFET›s transconductance is anywhere between.
   (a) 100 to 500 mA/V (b) 500 to 1000 mA/V
   (c) 0.5 to 30 mA/V (d) Above 1000 mA/V

17. A JEFT›s source terminal is analogous to a vacuum tube›s

(a) Plate (b) Cathode

(c) Grid (d) None Of The Above

18. A JFET›s output characteristics are very similar to those of a......... valve.

(a) Pentode (b) Tetrode

(c) Triode (d) Diode

19. In n-channel JEFT, an increase in the channel cross-sectional area results in a higher drain current.

(a) Is Increased (b) Is Decreased

(c) Remains The Same (d) None Of The Above

20. A JFET›s channel lies between the.............

(a) Gate And Drain (b) Drain And Source

(c) Gate And Source (d) Both A &B

ANSWERS															
1.	(a)	2.	(c)	3.	(b)	4.	(d)	5.	(b)	6.	(c)	7.	(d)	8.	(b)
9.	(a)	10.	(a)	11.	(b)	12.	(b)	13.	(b)	14.	(a)	15.	(d)	16.	(c)
17.	(b)	18.	(a)	19.	(a)	20.	(b)								

Short Answer Type Question

Category 1

1. Why is temperature co-efficient of resistance negative for semiconductor
2. Distinguish between intrinsic and extrinsic semiconductors.
3. Explain why diode is called as a unidirectional device.
4. Explain the current flow in a NPN transistor.
5. What is meant by biasing? Mention its types.
6. If both the emitter and the collector are made of the same semiconductor, why can't we switch people?
7. List the applications of Light Emitting Diode LED?
8. Define the transistor and draw the NPN transistor symbol.
9. Draw the NPN transistor's design.
10. Draw the symbol of N-channel & P-channel JFETs.
11. Create the Depletion type and Enhancement type MOSFET symbols.
12. State the application of JFET.
13. Compare FET& BJT.

Category 2

1. Draw & Explain N-channel JFET construction
2. Compare FET& BJT.
3. State different types of FET.

4. Distinguish between avalanche and Zener breakdown
5. Draw & Explain N-channel JFET construction
6. Explain the testing procedure of a regular pn-junction diode.
   How this is different from testing of zener diode
7. Draw the construction and explain testing procedure of MOSFET>
8. Explain testing of N-channel JFET and depletion MOSFET.
9. What factor should be considered while selecting equivalent of a transistor
10. Write short notes on SCR,BJT,OP-AMP,THYRISTOR

REFERENCES AND SUGGESTED READINGS

1. Gangopadhyay (1980), Production Technology-HMT, Tata McGraw-Hill Publishing Co. Ltd., New Delhi. ISBN 0-07-096443-2.
2. Gottlieb, I.M. (1985), Power Supplies; Switching Regulators, Invertors and Convertors, PBP Publications, Delhi.
3. Govindarajalu, B. (1991), IBM PC and Clones: Hardware, Troubleshooting and Maintenance, Tata McGraw-Hill Publishing Co. Ltd., New Delhi.
4. Harper, C.A. (1977), Handbook of Components for Electronics, McGraw-Hill Co., USA, ISBN 0-07-026682-4
5. Haskard,M.R.(1997), Electronic Circuit Cards and Surface Mount Technology, Technical Reference Publications Ltd., U.K. ISBN 1-872422-05-5.
6. Herrmann, G. and Egerer, K. (1992), Handbook of Printed Circuit Technology,
7. Electrochemical Publications Limited, U.K. ISBN 3-87480-056-3.
8. Khandpur, R.S. (1996), Handbook of Analytical Instruments, Tata McGraw- Hill Publishing Co. Ltd., New Delhi. ISBN 0-07-460186-5.
9. Khandpur, R.S. (1987), Troubleshooting of electronic equipment McGraw-Hill Publishing Co. Ltd., New Delhi.McGraw-Hill, U.S.A. ISBN 0-07099479-X.
10. https://nptel.ac.in/courses/115102025
11. https://www.youtube.com/watch?v=F5IpIHzBvLQ
12. https://www.youtube.com/watch?v=h0Y9jDKqScQ&list=PLgMDNELGJ1CaNcuuQv9xN07Z
13. https://www.youtube.com/watch?v=Fjm9Nocs0Aw

KNOW MORE

Wafers are highly purified silicon slices that attend as the beginning substrate for nearly all chips.Between etching, stripping, and cleaning this method can have up to and over 500 various steps to build the wantedchip. That's about 140 chips for any person on the planet, or 130 million per sixty minutes.If all these semiconductors were stacked one on top of an additional they'd extend above the greatest cruising altitude for commercial aircraft.

QR SCAN CODE FOR SUPPORTIVE KNOWLEDGE EMBEDDED IN THE CHAPTER

□□

4 IC Testing

UNIT SPECIFICS

The following topics have been explained in this unit:

- Understanding the types of Digital IC
- Explain Digital Integrated troubleshooting method
- Describe Testing of flip-flop and digital IC.
- Explain Working principle of digital device like logic Probe and logic clip,
- Knowledge of logic comparator and logic current tracer and other logic devices
- Understanding the concept of de-multiplexers ,registers and counters, multiplexers and encoders, decoders

The study of these issues' real-world applications is meant to inspire more inquiry and originality, as well as to sharpen the ability to solve problems. The unit provides a significant number of multiple choice questions, questions with short and lengthy answers separated into two groups based on the lower and higher orders of Bloom's taxonomy, a list of references, and suggested readings that can be utilised for practise. It's worth noting that QR codes, which may be scanned for further information, have been included in various parts so that readers can learn more about the themes that interest them.

There is a "Know More" section depending on the information that comes after the associated practical. The supplemental material included here has been carefully organized to maximize its usefulness to readers. This segment specializes in the beginning activity, examples of some captivating facts, analogies, the records of the subject's evolution at the same time as emphasizing considerable observations and findings, timelines beginning with the introduction of the relevant subjects as much as the prevailing, packages of the issue rely for our real existence and industrial packages on a variety of factors, and case research each day environmental, sustainability, social, and moral issues

RATIONALE

This unit on IC testing helps students to get knowledge about integrated circuits and their testing .It explains the Packages in Digital Integrated Circuits, Digital troubleshooting methods- This unit will also explain various concept about the identification of integrated circuits, integrated circuit pin-outs and methodology to handling integrated circuits,

PRE -REQUISITES

Physics: Integrated circuits ,Basic knowledge about digital circuit.

UNIT OUTCOMES

This unit's outcomes are shown below.:

U4-O1: Learn how integrated circuits function and how to recognize them.

U4-O2: Describe digital troubleshooting methods and fault diagnosis in digital circuits

U4-O3: Study the Logic clip and Logic Probe, Current tracer and Logic pulser, Logic comparator

U4-O4: Explain handling precautions for ICs

U4-O5: Describe testing counters, registers flip-flops, multiplexers and encoders and decoders, De multiplexers

Unit-4 Outcomes	EXPECTED MAPPING WITH COURSE OUTCOMES (1 - Weak Correlation; 2 - Medium Correlation; 3 - Strong Correlation)					
	CO-1	CO-2	CO-3	CO-4	CO-5	CO-6
U4-O1	1	1	1	2	2	1
U4-O2	2	2	2	3	2	2
U4-O3	3	3	3	2	3	1
U4-O4	2	1	2	3	1	2
U4-O5	3	3	3	-	2	2

4.1 INTRODUCTIONOF DIGITAL INTEGRATED CIRCUITS

Digital IC are substantially used in almost fields of electronics from computing to industries management,digital devices, verbal exchange systems, and clinical device. It seems that digital circuits will eventually be employed in every face of electronics. There is every system is run by integrated Circuits. There is the option of purchasing electronic components in either a bundled or a separate format. The transistor, diode, resistor, capacitor, and inductor are examples of discrete components. That electronics are as ubiquitous as they are now is largely due to ICs. The term "integrated circuit" (IC) refers to a device that can contain both active and passive components, as well as any type of circuit, including analogue, virtual, and mixed-signal circuits. Everything from ultra-high-tech corporate cars to microprocessors uses digital circuitry. However, logic gates are the most basic building blocks of digital circuits, therefore understanding their basics is essential for developing proficiency in tedious process.

Know more about Details IC

4.2 LOGIC IC FAMILIES

The essence of the simple logical component relies upon the characteristics of electrical elements applied to realize them. In the past era diodes have been largely used as a digital technique, for developing any circuit. And it changed into obvious to take the AND or OR gates due to the fact the fundamental factors. When transistors finally came into their own, basic logic circuits based on NAND and NOR gates were a no-brainer. the reason for this is because the transistor's output signal is inverted with respect to its input signal. Transistor-transistor logic and complementary metal-oxide semiconductor logic family, the most well-known and widely utilized circuits in modern digital systems, are becoming problematic. With the advancement of integrated circuits, the issue of size has been resolved, and it is now possible to acquire a wide variety of functions on a single chip. It's fair to say that every generation of IC gate shares the same logical qualities and physical properties.

4.2.1 Transistor-Transistor Logic (TTL)

The transistor-transistor family is the most powerful and widely used type of logic device. Semiconductors allow for the creation of the many logic gates inside the IC. In TTL circuits, the bipolar transistor plays a vital role. The input devices used in TTL generation all have multiple emitter transistors. Totem pole output circuits are used in TTL gates. All other types of output circuits use an open collector output, which necessitates an extra pull-up resistor for determining the exact high and low values of logic outputs. TTL logic is widely used because of its low cost, high speed (especially the Schottky TTL variant), ease of interfacing with other digital circuits, and fan-in/fan-out capabilities. TTL logic's main characteristics include a propagation latency of 10 ns, a flip-flop charge of 20 MHz, a fan-out of 10, a noise margin of 0.4V, and a power dissipation of 10 mW. TTL gates of the 74 series are rated for operation up to 70 degrees Celsius. The 54 series, on the other hand, may be safely used up to 125 degrees Celsius. There is typically more than one gate in an IC package. TTL stands for transistor-transistor-logic, and there are many different kinds of TTL families.

4.2.2 Schottky TTL

Schottky TTL logic gate are faster and popular TTL because it uses less power. The 74S/54S series. are available in the integrated form and 54LS/74LS. series is also commercially available as low power Schottky TTL series

4.2.3 Emitter Coupled (ECL) Logic

The emitter coupled logic family offers different method for the gate to accomplishing better speed, it makes this logic family better from the other forms of logic families. Emitter-coupled logic is the best BJT-based logic family used in the conventional logic-system design. Sometimes, it is also called current mode logic which is a very high-speed digital technology. Generally, ECL is considered as the fastest logic IC where it achieves its high-speed operation by using a very small voltage swing & also avoiding the transistors from entering the saturation region.

4.2.4 Complementary Metal Oxide Semiconductor Logic families

The CMOS family offers numerous benefits over bi-polar transistor-based logic circuits, namely low power dissipation and adequate noise tolerance. As opposed to simple gates and flip-flops,the method is more appropriate for the creation of large-scale integrated circuits. As a major plus, CMOS technology allows for a wide array of components to be packed in very close together. CMOS gates that are obtainable commercially belong to the 4000 series. For example, a quad2 input AND gate and quad2 input NOR gate in CMOS are represented by the numbers 4081(7408TTL) and 4001,respectively.

4.3 PACKAGES IN DIGITAL IC'S

A variety of packaging is used for the digital or electronic components. Numerous factors are decided by the packaging. If the problem can be used on a PCB, prototyping board, or both is the first question. There are a variety of IC packages available for digital additives to accommodate various product lifecycle stages (including prototype and manufacturing) and address a wide range of circuit design problems. When using an icon a PCB, the packaging is very crucial IC packages come in a wide variety of forms, and they can be categorized using a variety of techniques. the heart of any IC is made of semiconductor wafers, which can be layered with copper and other materials in a complex configuration materials. A die is the reduced and shaped assembly of those semiconductor wafers. All of its are in the die in the final stage of IC production, microscopic-scale additives and interconnections are created. So that the IC can connect to various components on a PCB, microscopic electric contacts are removed from the die encasing the semiconductor "die" and eliminating macroscopic electric contacts are the main goals of IC packaging connections on a PCB.

Three purposes are specifically served by the packaging.

1. It shields the semiconductor circuit.
2. Harm or physical impairment. It guards against corrosion on the circuit.
3. The PCB over which the semiconductor device's electric contacts are arranged. that is a crucial consideration for each IC design during the PCB designing process.

4.3.1 IC Packages Classification

Following are the manner in which IC programs classification relies upon

1. IC package systematically classified consistent with their mounting vogue. Mounting style is one among the conventional intensive considered integrated circuits. All IC package lies in huge categories by the method of mounting style like PTH suggest that plated through-hole and SMT suggest that surface-mount.
2. The placement of the IC's pins on the package can also serve as a further categorization criteria. Every IC is a standardized square or rectangle or long thin strip. To further categorize the uses of ICs, pin shapes will be used.The leads are given in following shape
   - Linear
   - Mutual folding
   - Needle-shape
   - L-shape
   - Tape or Movie type
   - J-shape
   - Electrode bud etc.

3. IC packages are familiar with the terminal pin. There are two-terminals, 3-terminals, four-terminals, 5-terminals, six-terminals, and over six-terminals IC applications the dimensions of the terminals conjointly is a distinctive considering similar pack types.
4. IC packages are with the help of terminal mater. There are terminal three terminals, four terminals, five terminals, or over half-dozen terminals packages of IC. The dimension of the terminals conjointly is a necessary side in a comparable different package.

Following are the package sorts

1. Single inline package
2. Zigzag inline package
3. Dual inline package
4. Quad inline package
5. Ceramic flat package
6. Surface mount-small outline package
7. Surface mount lead-less package
8. Flat pack package
9. Chip carrier package
10. Chip scale package
11. Grid Matrix or array package

1. Single inline packages

Single inline packages have one row of pins and are mounted via a hollow centre. These wires, along with the rest of the boundary-line package, are stacked neatly in a vertical orientation. These packages aid in reducing PCB costs by eliminating the need to decrease the wire pitch, which is a common method for doing so here (Shriveled Single Inline Package) SSIP, (single-in-line packages) SIP, and (Single Inline Package with Heat

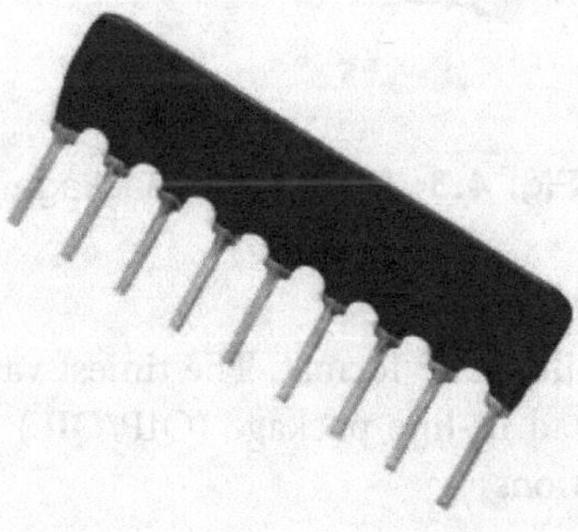

Know about logic circuit

Fig. 4.1: SIP Package

2. Zigzag Inline

The pins in these packages are all located in a single vertical row. This mounting method utilizes holes. The leads are laid out in a vertical fashion along the package's outside edge, in a fashion very dissimilar to SIP but with a zigzag pattern instead of mutual folding. The mostly used zigzag inline package are ZIP and SZIP means small zigzag In-line package

Fig. 4.2: Zigzag inline package

3. Dual In-line Packages

Through-hole mounting is standard for these dual in line packages, which also have rows of pins. The wires run in two parallel rows along the length of the package. In the realm of packaging, this is a well-known pattern. As a default, DIP is one of the most common IC package types for devices with 6–40 pins. these ICs may be offered in surface-mount packages for industrial applications. Using dual in-line packages is easy since they can be used with common components found on breadboards and prototype boards. Common types of dual-inline packages include SDIP(Shrink Plastic Dual In-line Package),CDIP or CER-DIP(Ceramic Dual In-line Package), (Plastic Dual In-line Package) PDIP, (Skinny Dual In-line Package) SKDIP.

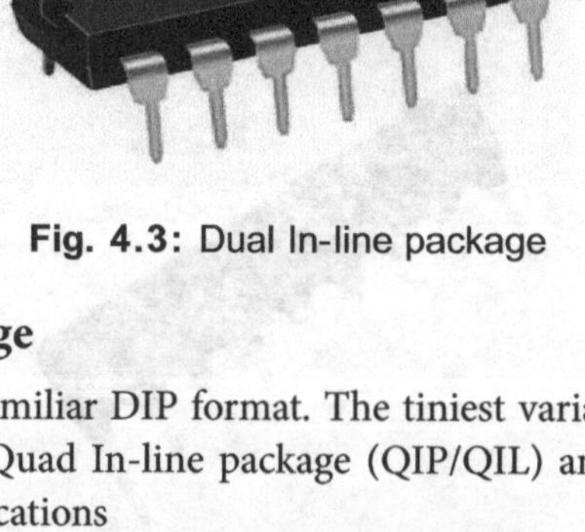

Fig. 4.3: Dual In-line package

4. Quadruple Inline Package

This QIP is very much like the familiar DIP format. The tiniest variation is that some rows' leads are slanted and others' are straight. Quad In-line package (QIP/QIL) and ceramic lead significantly less often than other quad inline applications

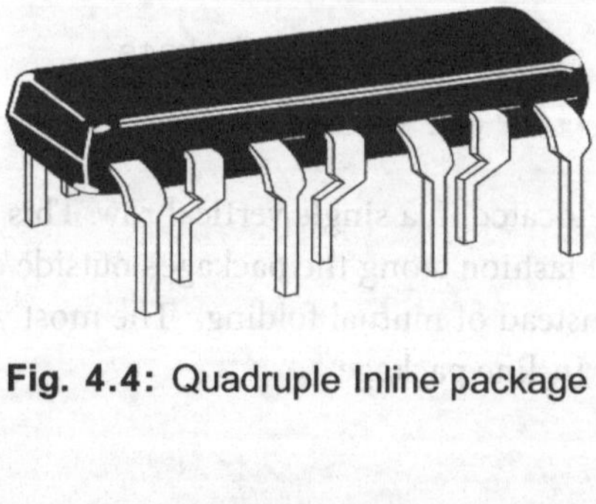

Fig. 4.4: Quadruple Inline package

5. Ceramic Flat Pack

These IC packages were developed for united State military PCB standards, which need a narrower pitch (50 mils). When using a surface-mount technique, the wires are laid out in two or four parallel rows across the board. These packages have narrower pitches, which necessitates more delicate IC handling and more costly board processing. Due to prohibitive PCB prices, these packages are seldom used in commercial settings. Ceramic flat pack and quad ceramic quad flat pack (QCFP) are two typical ceramic flat pack packaging

Fig. 4.5: Ceramic Flat Pack

6. Surface Mount Small Outline Packages

Two rows of terminals and a surface-mount mounting technique distinguish the small shape surface mount package variants. Terminals might be L-shaped, J-shaped, or lead-free. These packages are commonly utilized by SMD components. To minimize the amount of space needed, J- leaded packages were developed. Electrode pads serve as the connections' terminals in lead-free models. the followings are main small-outline surface-mount packages.

- Small outline (SOP) Package
- Ceramic small outline (CSOP) package
- Dual small outline (DSOP) package)
- Thermally enhanced small outline (HSOP) package
- Shrink Small outline SSOP) package
- Thin-Small Outline (TSOP) package
- Thin Shrink Small Outline package
- Thin very small outline (TVSOP) package
- Mini or Micro Small Outline(MSOP) Package
- Thermally-enhanced shrink small-outline (HSSOP) package
- Thermally-enhanced thin shrink small-outline (HTSSOP) package
- Quarter Small Outline(QSOP) Package
- Small Outline Integrated Circuit (SOIC)
- Small Outline Integrated Circuit Wide (SOICW)
- Plastic small-outline (PSOP) package
- Plastic small outline no lead (PSON) package
- Very-small outline (VSOP) package

- Very shrink small outline package(VSSOP)
- Small Outline J leaded (SOJ) package
- Small Outline Non leaded (SON) package

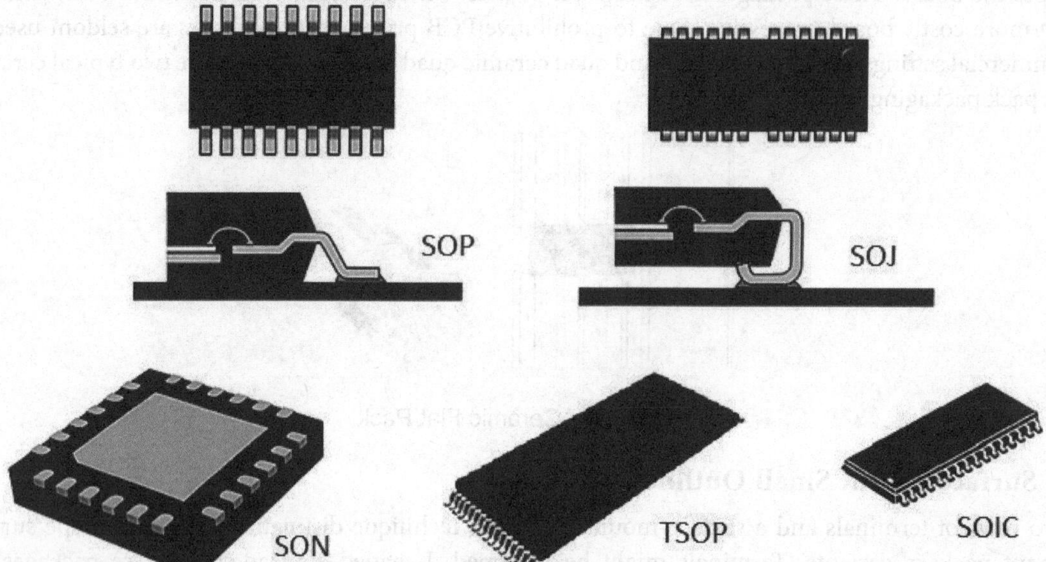

Fig.4.6: Surface Mount Small Outline

7. Surface Mount Lead-less Packages

There are no leads on these surface-mount IC packages. The connection terminals are the electrode pads along each of the four edges. There may be other categories into which the surface mount lead- free packages can be divided.

- LCCs are the most prevalent kind of lead-free surface-mount package (Lead-less Chip Carrier)
- PQFP (Plastic Quad Flat Pack)
- PLCC (Plastic Leaded Chip Carrier)

8. Flat Pack

Two or four rows of terminals surround the IC's perimeter in these IC packages. L-shaped, J-shaped, and lead-free terminals are utilized for surface-mount components..

The following are examples of typical flat pack IC packages:

1. Quad Flat (QFP) Package
2. Thin Quad Flat (TQFP) Package
3. Small Thin Quad Plastic Flat (STQFP) Package
4. Fine Pitch Quad Flat (FQFP)Package
5. Quad Flat Package with Heat Sink(HQFP)
6. Low profile Quad Flat (LQFP) Package

7. Very small Quad Flat (VQFP) Package
8. Metric Quad Flat (MQFP) Package
9. Bumper Quad Flat (BQFP)Pack
10. Exposed Thin Quad Flat (ETQFP) Package
11. Power Quad Flat (PQFN) pack
12. Plastic Quad Flat (PQFP) Package
13. Quad Flat J Leaded (QFJ) Package
14. Quad Flat Non leaded (QFN) Package
15. Thin Quad Flat No Lead Plastic(TQFN) Package
16. Dual Flat (DFN) Package
17. Quad Flat Leaded (QFI) Package
18. Heat Sink Very Thin Quad Flat Pack No Leads (HVQFN)

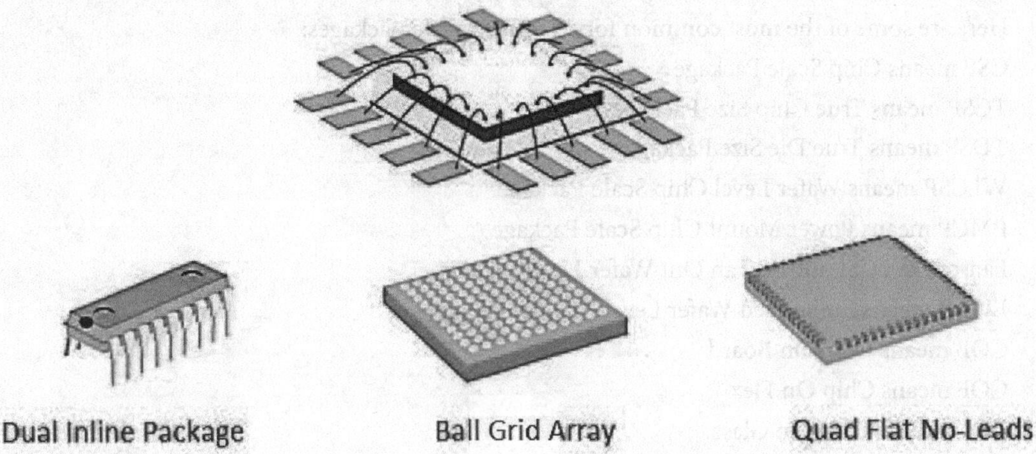

Fig. 4.7: Some IC related to Flat pack

9. Chip Carrier

Terminals are located on all four sides of these square chip carriers, sometimes known as chip career IC packages. like J-leaded and lead-free terminals . The most popular chip carrier packaging types are as follows

- BCC' (Bump Chip Carrier)
- 'LCC' (Leaded Chip Carrier)
- 'LCCC ' (Leaded Ceramic chip Carrier)
- 'PLCC' (Plastic Leaded Chip Carrier)
- 'LCC'(Lead less Chip Carrier)
- 'CLCC' (Ceramic Lead-less Chip Carrier)
- ' DLCC' (Dual Lead-less Ceramic Chip Carrier)

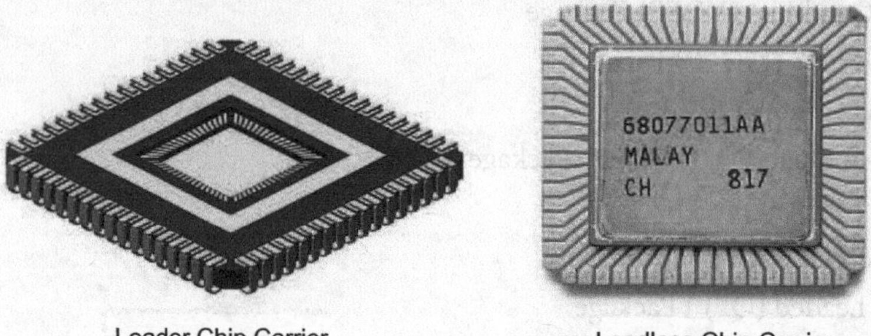

Fig. 4.8: Chip Carrier

10. Chip Scale/Non-packaged

These IC packages are noticeable due to their size, which is practically identical to that of the surrounding silicon wafer and the term "non packaged ICs" is another name for them.

- Here are some of the most common forms of chip scale packages:
- CSP means Chip Scale Package
- TCSP means True Chip Size Package
- TDSP means True Die Size Package
- WLCSP means Wafer Level Chip Scale Package
- PMCP means Power Mount Chip Scale Package
- Fan-out WLCSP means Fan Out Wafer Level Package
- EWLB means Embedded Wafer Level Ball Grid Array
- COB means Chip On Board
- COF means Chip On Flex
- COG means Chip On Glass
- COW means Chip On wire
- TAB means Tape Automated Bonding
- MICRO SMD

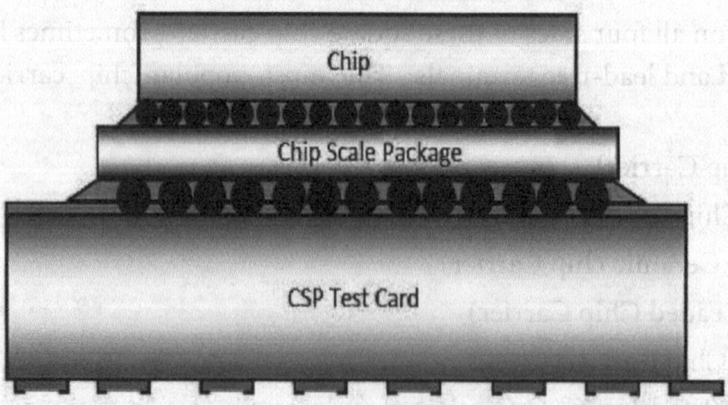

Fig. 4.9: Chip scale package

11. Grid Array

These IC packages have a variety of terminals and pins underneath and are square or rectangular in shape. The terminals may be as follows

- SMTBGA Means Surface mount type with solder Ball Grid Array
- BGA means ' Ball Grid Array'
- EBGA means ' Enhanced Ball Grid Array'
- EWLB means ' Embedded Wafer Level Ball Grid Array'
- FTBGA means ' Flex Tape Ball Grid Array'
- TFBGA means 'Thin and Flange Ball Grid Array'
- OPGA means Organic Pin Grid Array
- FCPGA means Flip Chip Pin Grid Array
- MAP BGA means Mold Array Process Ball Grid Array)
- UBGA means Ultra Ball Grid Array
- TBGA means Thin Ball Grid Array
- SBGA means Super Ball Grid Array
- UFBGA (Ultra-Fine Ball-Grid Array))

4.4 IDENTIFICATION OF IC'S

The dual inline (DIP) package is typically used for digital integrated circuits.

- Identification of a specific IC because sometimes device could be an analog component, such as an operational amplifier etc. One of the two ways listed below is used to represent ICs in a schematic diagram.
- The rectangle in Figure 4.9, represents IC with pin numbers displayed on each pin. On the schematic, the IC'sidentification number is listed.

(a) For instance, the quad 2-input AND gate IC74 LS 08 is denoted in a schematic as 1/474LS 08 (Fig.4.10) information provided on the IC itself can be used to identify an IC. Despite being standardized, the numbering system varies according to manufacturer . The surface of an IC having markings as shown in Fig. (4.11). The logic families and their functions are identified by the Core number. The first two numbers, "74," indicate that the integrated circuit (IC) in question belongs to the 7400 series IC family. The IC's purpose is denoted by the last letters. Letters put into the middle of the core number indicate the subfamily of logic being used. When writing the popular TTL series, no letters are added to the main number.

(b) The identically numbered Integrated Circuit's in every family have the exact same pin rs and carry out the same functions. But due to variations in timing and power requirements, they cannot be interchanged. A prefix before the core number denotes the manufacturer. The core number's suffix provides information about the package type, temperature range, etc. (Fig.4.11)

(c) The production batch number and year of manufacture is also offered in some ICS. For instance, the code 8234 signifies that the device discovered in 1982, in the thirty four batch. Along with other details regarding the IC, the manufacturer's trademark (logo) is also printed.

(d) Information can be found from the most popular series, like 7400, from a variety of sources. Truth tables, pin-outs,and other information are all provided in data books .

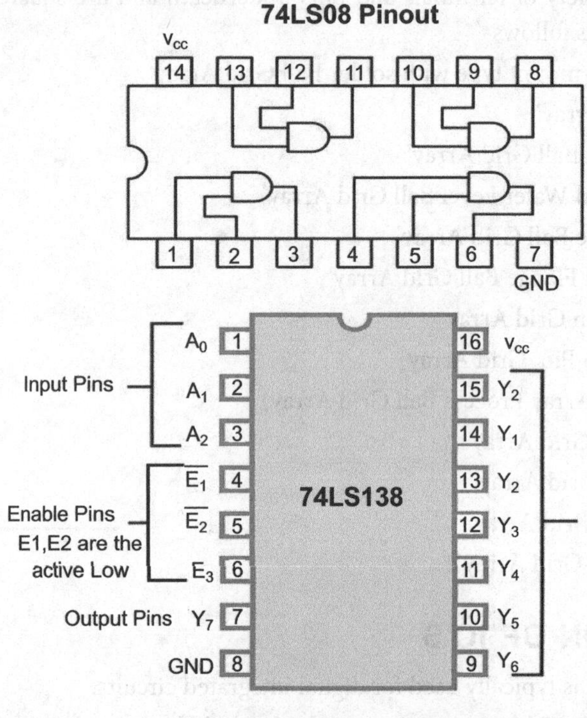

Fig. 4.10: Representation for a digital IC

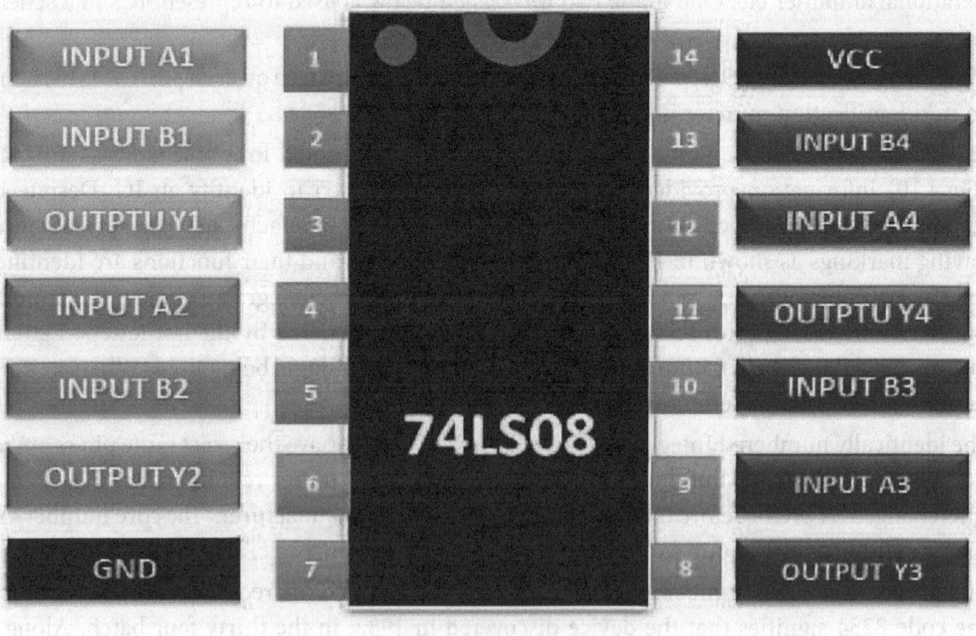

Fig. 4.11: Representation of IC74LS08 pins

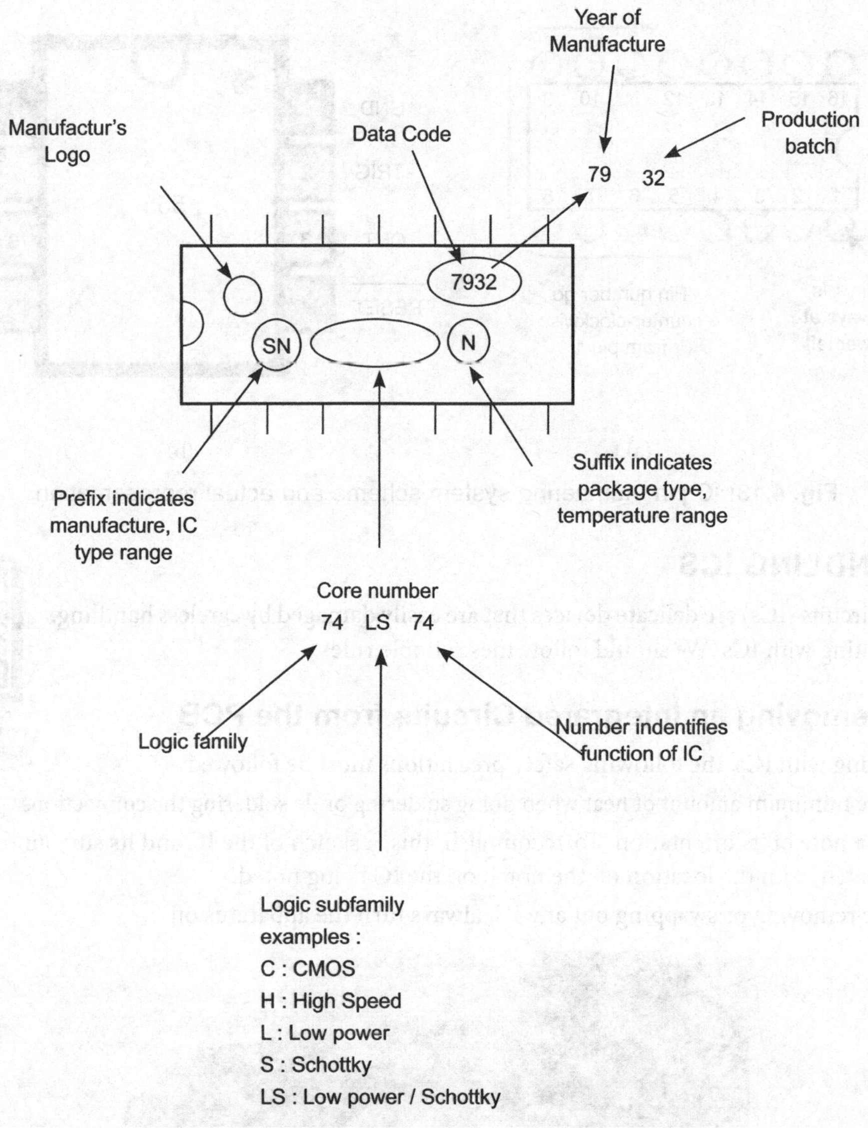

Fig. 4.12: Identification marks on the digital ICs

4.5 IC PIN-OUTS

A pin connection diagram that depicts the signals is part of the IC's technical information.And gives information about each Pin connections (as shown in Fig.4.12) . A dot mark is used to identify the IC's one end and plastic notch placed on it. On the end of the IC, pin 1 is always located in the upper left corner with incorporates the notch. The numbers are displayed on the IC in a row, going from left to right pin. The supply voltage pin on the connection diagram is denoted by V. This voltage in most ICs is 5Vdc.Ground pin on the diagram denotes the pin connection for ground. while GND is the pin with the lowest value. However, this is not always true.

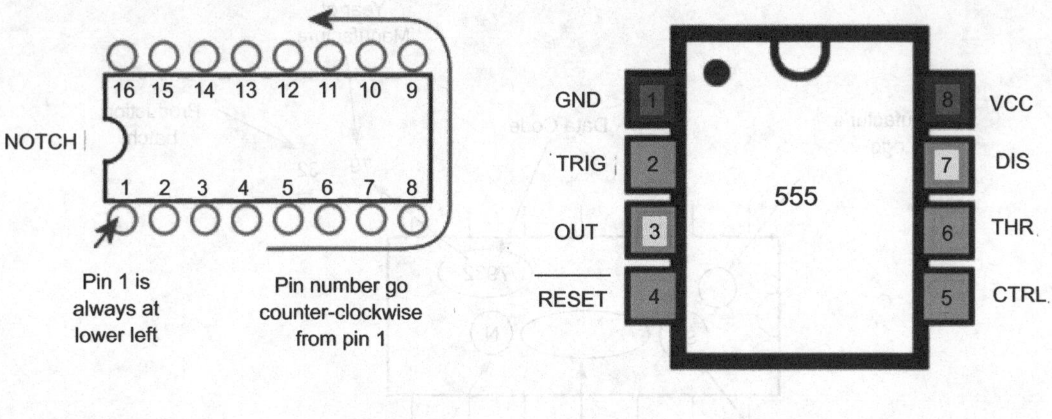

(a) (b)

Fig. 4.13: IC pin numbering system scheme and actual representation

4.6 HANDLING ICS

Integrated circuits (ICs) are delicate devices that are easily damaged by careless handling. When operating with ICs .We should follow these simple rules

Know more about Combi national circuit

4.6.1 Removing an Integrated Circuits from the PCB

When working with ICs, the following safety precautions must be followed:

(a) Use the minimum amount of heat when doing soldering or de soldering the connections

(b) Make a note of its orientation. To accomplish this, a sketch of the IC and its surroundings should be created, with the location of the notch on the IC being noted.

(c) Before removing or swapping out any IC, always turn the apparatus off

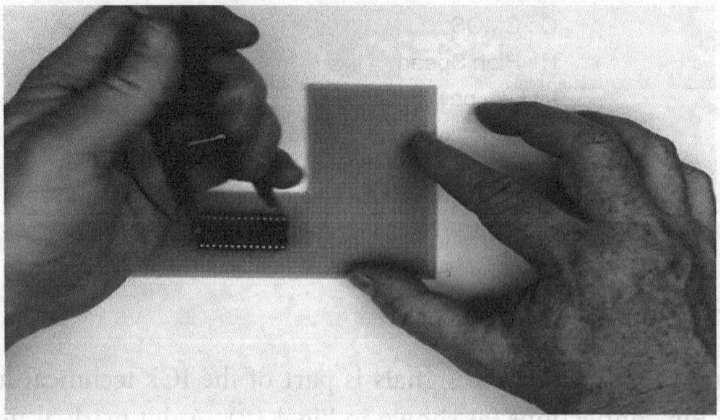

Fig. 4.14: DIP insertion-extraction clip and IC Remover

4.6.2 Mounting integrated circuit in the circuit board

Make sure the Integrated Circuit's pins are straight if it is going to be installed in the socket. A few insertion tools include a tool for straightening the pins. (Fig. 4.14) illustrates Pin rows should be inserted one at a time, then the other, Make sure to position the IC with the notch correctly. When installing the IC socket, the IC must be removed from the socket. This ensures that ICs will survive the soldering process without being damaged by the heat. Use a flux remover to clean the back of the circuit board after installation the socket. It's possible that any flux residue will be sticky and attract dirt. A minor leak brought on by the dust might create the impression that the IC is damaged.

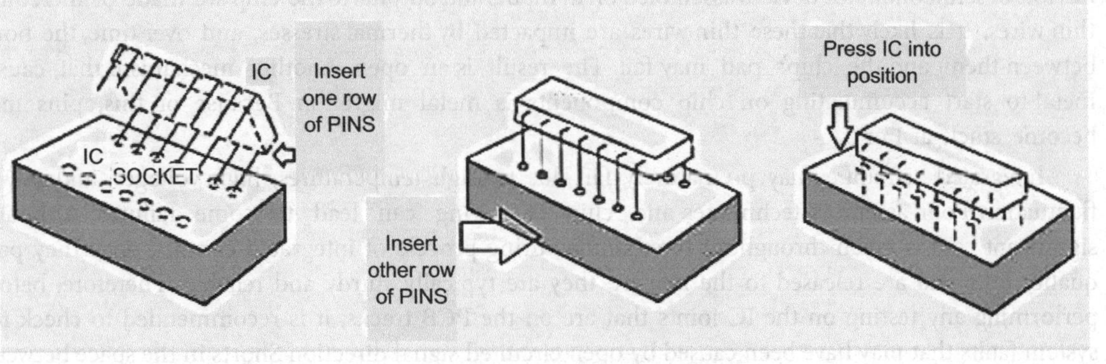

Fig. 4.15: Mounting IC in a socket

4.6.3 Lifting an IC Pin

When the Integrated Circuit is mounted on a socket, many issues may be solved by removing one/two pins from the circuit board. after removing pins intermediate any one pin may be bend . So that it will not be in the socket's path. this bent pin where it emerges from the packaging using long-nosed pliers (Fig. 4.15). Only a 45° bend will be necessary for it when it will be pass through the socket. When flip-flops stop working, you often have to pull it up on many leads simultaneously. Some inputs may need to be tied HIGH or LOW when they are released

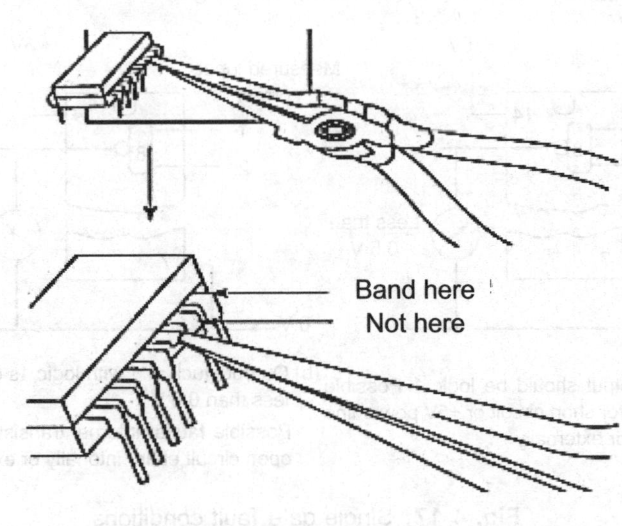

Fig. 4.16: Use of a plier for lifting IC

4.7 DIGITAL TROUBLESHOOTING METHODS

4.7.1 Typical Faults

The critical issue for chip manufacturers is getting the input and output signal with a small chip, which has lots of semiconductor devices assembled on it. Inputs and outputs to the chip are made of incredibly thin wires. It is likely that these thin wires are impacted by thermal stresses, and over time, the bond between them and the chip's pad may fail. The result is an open. Another mechanism that causes metal to start accumulating on chip components is metal migration. Because of this, pins may become stuck at 1 or 0.

Integrated circuits may prematurely fail due to high temperatures, high voltages, and power fluctuation Poor assembly techniques and chip packaging can lead to some failures. Although significant care is taken throughout the manufacturing process of integrated circuits, once they pass quality tests and are released to the market, they are typically sturdy and reliable. Therefore, before performing any testing on the IC joints that are on the PCB tracks, it is recommended to check for system faults that may have been caused by open circuited signal direction Shorts in the space between the PCB tracks failure of discrete components located outside of the IC.A An internal IC failure could be the root cause of any such malfunction. A connection between two pins that is not grounded or at the V_{AC} potential may have any of the following issues: an open lead at an input or output pin; a connection between an input terminal and the ground or to a power source;a connection between an input terminal and ground or to a power source when the IC's internal circuit fails. Possible fault scenarios are shown in Figure 4.16 (a) for a single gate. It explains that with logic l's inputs the output of 4.16 (b) remains at 1, even though it should be less than 0.8 V. Internal open circuits in the 0V line or internal open circuits in transistors are two potential faults. The 0V line is considered to be externally sound. In (Fig. 4.16(b)) It is at logic 1, but in (Fig.4.16(a)), the output is stuck at 0'.An internal transistor that has been shorted out or an internally open V_{AC} line are two potential faults in this gate.

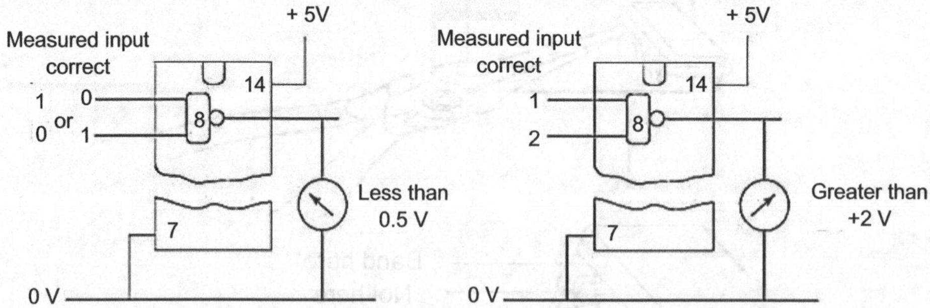

(a) Output stuck at 0, output should be logic 1 Possible faults : Internal transistor short circuit or +5V power line open either internally or externally

(b) Output stuck at 1 with logic 1s on inputs-output should be less than 0.8 V

Possible faults: Internal transistor open circuit or 0 V line open circuit either intenally or externally

Fig. 4.17: Single gate fault conditions

An IC generally experiences a fault that is stuck at at same value which obstructs the signal flow, when its internal or steering circuitry fails.The failure's internal or external nature can only be determined by conducting a investigation of the circuit.To summarize, (Fig. 4.17) displays typical issues with digital integrated circuits. These include shorted gate to +VC

- Open gate
- Uncut pin connection
- Quick travel from one terminal to another
- Grounding gate shorted
- gate length reduction

Know more about Logic probe

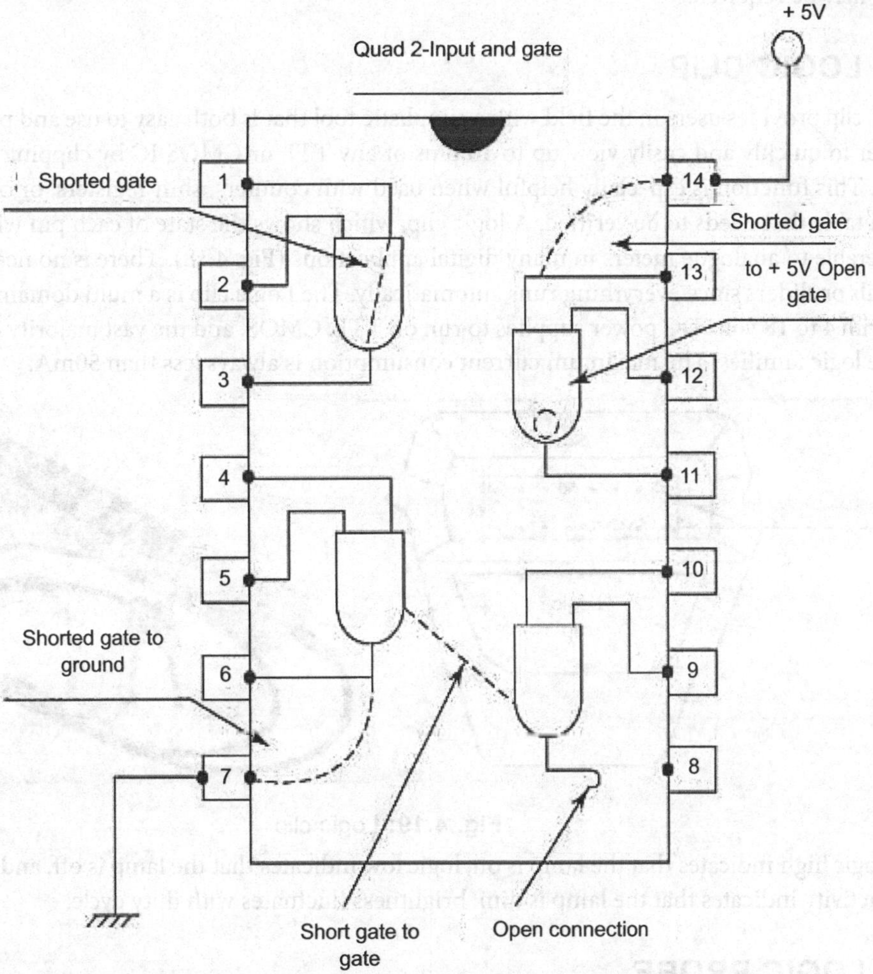

Fig. 4.18: Common faults encountered in digital IC

4.7.2 Testing Integrated Circuits with Pulse Generator

On logic integrated circuits (ICs), three different tests can be run. DC and AC functional tests are being conducted. While DC and AC tests are more frequently conducted in production and functional testing is the test that is most frequently used in servicing.

The practical test merely concludes whether the IC working according to its "truth-table". During this test, an IC is stepped through its numerous states using a sequence of pulses from a pulse generator. The outputs of the IC can be connected to an oscilloscope or a logic monitor in order to observe those states. The output of the pulse generator must be adjusted such that it corresponds to the generator's baseline, which should be at 0 volts, and the pulse amplitude, which should be set considerably above the minimal for logic 1. Most combinational logic integrated circuits (Gates, Inverters, Decoders, Encoders, Multiplexers, etc.) may be tested in a functional manner. If functional testing shows that an integrated circuit is behaving according to its truth table but the circuit is still not functional, then DC parameter testing may be required.

4.8 LOGIC CLIP

A logic clip provides users in the field with a simplistic tool that is both easy to use and portable.It allows the user to quickly and easily view up to 16 pins of any TTL or CMOS IC by clipping on to any such device. This function is especially helpful when used with counters, shift registers, or circuits that have a truth table that needs to be verified. A logic clip, which shows the state of each pin with its own LED, is preferable to analogue meters in many digital applications (Fig. 4.18). There is no need to fiddle with any dials or sliders since everything runs automatically. The Logic clip is a multi domain clip that can be used with 4 to 18 volts DC power supplies to run on TTL, CMOS, and the vast majority of other positive voltage logic families. The maximum current consumption is always less than 50mA.

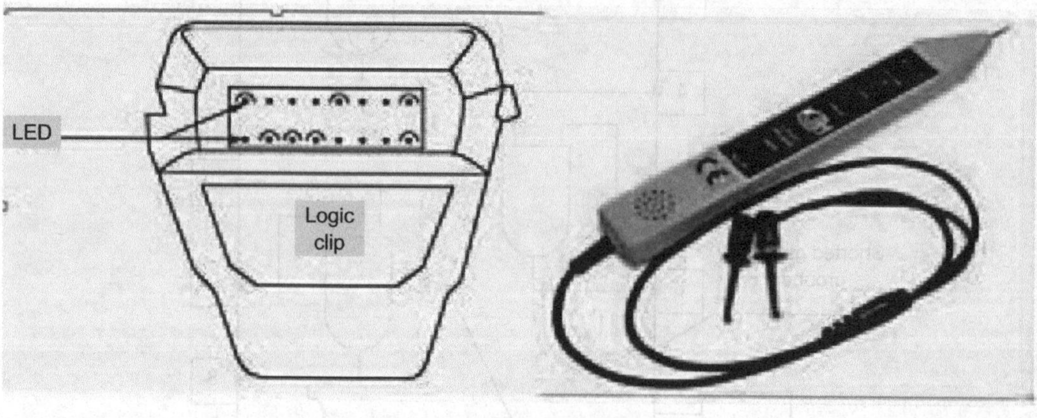

Fig. 4.19: Logic clip

Logic high indicates that the lamp is on, logic low indicates that the lamp is off, and pulse activity indicates that the lamp is dim; brightness fluctuates with duty cycle.

4.9 LOGIC PROBE

A logic probe is a low-cost hand-held test probe used for analyzing and troubleshooting the logical states of a digital circuit. On the body of the probe, the majority of contemporary logic probes often have one or more LED a high (1) logic state LED indicator. a low (0) logic condition can be indicated by an LED. an LED to show the transition between low and high states. A pulse-stretcher circuit is typically a part of the pulse-detecting electronics, allowing even extremely brief pulses to be seen on the LED. A control on the logic probe enables either continuous operation or the collection and storing of a single event. None of the LED illuminate when the logic probe is either linked to an erroneous logic level (a fault condition or a tri-stated output or is not attached at all.

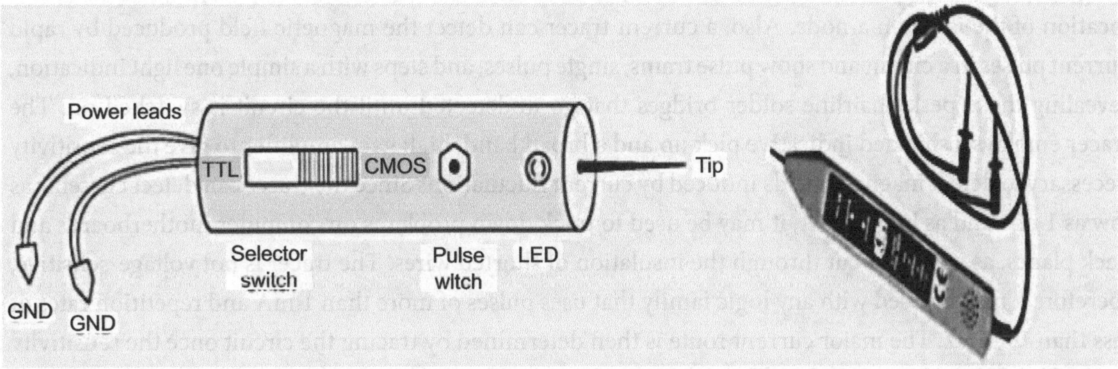

Fig. 4.20: Logic probe

A logic probe's ability to stretch pulses, slowing and lengthening them at the display so the operator can easily perceive them, is one of its most useful capabilities. The logic probe's optional pulse memory unit records and visualize short pulses that would otherwise be hidden. Any pulse that the probing tip detects is stored in memory and displayed until the RESET button is pressed. A red LED for logic 1 and a green LED for logic 0 are included in some logic probes, respectively. There are several possibilities available with these probes.

4.10 LOGIC PULSER

The logic pulser is yet another extremely useful tool for diagnosing logic circuit issues. It is made to inject a logic pulse into the circuit being tested and has a shape that is similar to the logic probe. Most of the time, logic pursers are used in conjunction with a logic clip or a logic probe to help you trace the pulse through the test circuit or check that an integrated circuit is working properly. A feature on some logic pursers allows for either the injection of a single pulse or a train of pulses. Logic pursers are typically powered by an external dc power supply, but in some instances they may be directly connected to the device under test's power supply.

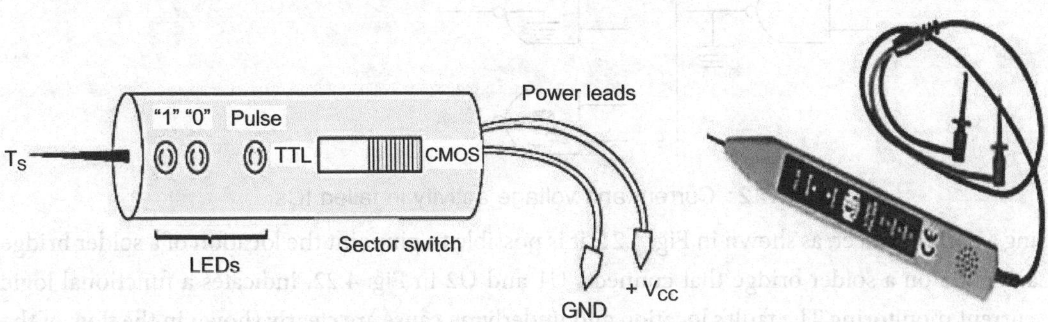

Fig. 4.21: Logic pulser

4.11 LOGIC CURRENT TRACER

A logic current tracer is used to locate low-impedance defects in digital circuits and pinpoint the exact location of the issue on a node. Also, a current tracer can detect the magnetic field produced by rapid current pulses in a circuit and show pulse trains, single pulses, and steps with a simple one light indication, revealing those pesky hairline solder bridges that go undetected until the circuit is switched on. The tracer employs a shielded indicative pick up and a broad band, high gain amplifier to give the sensitivity necessary to detect magnetic fields induced by current fluctuations Since the tracer can detect currents as low as 1 mA and as high as 1 A, it may be used to track down problems on computer motherboards and back planes, as well as to cut through the insulation of shorted wires. The tracer is not voltage-sensitive, therefore it may be used with any logic family that uses pulses of more than 1 mA and repetition rates of less than 10 MHz. The major current route is then determined by tracing the circuit once the sensitivity control has been adjusted to show the existence of AC current activity. When the tracer's end is taken off the primary current route, the lamp turns dark along the primary path but continues to shine along the secondary way. As you go from point to point or follow traces, the light intensity will shift; after you've located the fault, the tracer will likely be as brilliant as it was at the starting point.

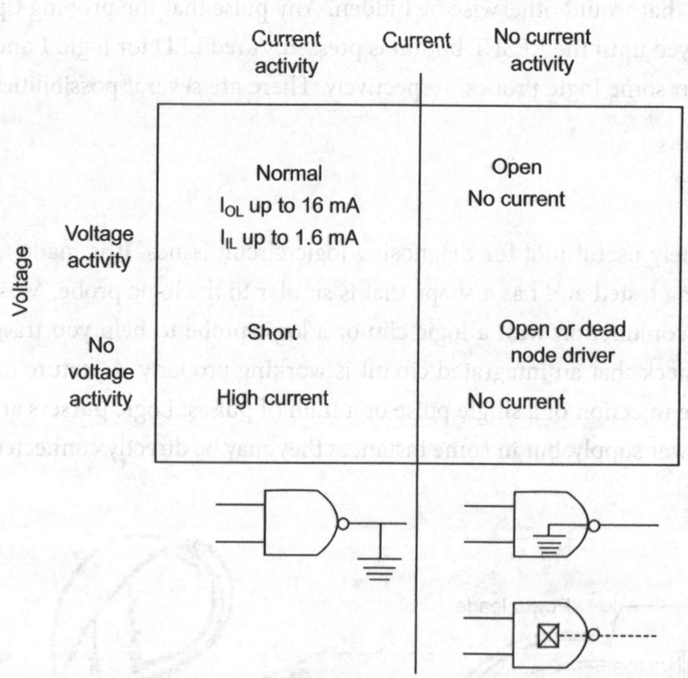

Fig. 4.22: Current and voltage activity in failed ICs

Using a current tracer, as shown in Fig. 4.21, it is possible to pinpoint the location of a solder bridge fault. Each node on a solder bridge that connects U1 and U2 in Fig. 4.22. indicates a functional logic failure. current monitoring The fault's location and underlying cause are clearly shown in the flow of the circuit. In this instance, the current that is being tested has just happened.

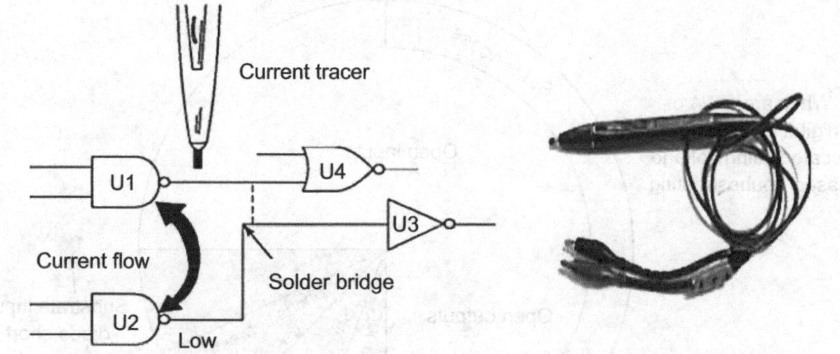

Fig. 4.23: Solder bridge fault detection using a current tracer

Multiple input fault case shown in Figure 4.23. Due to a short in Gate U5A, the node is permanently in the LOW state, and all inputs, including U 1, are effectively grounded. A simple, unambiguous, and single-lamp sign of current activity on the node is provided by a current tracer, allowing for fast verification of this problem. In this scenario, a logic pulser supplies the brief burst of current. Expert analysis of digital circuit failures has shown that input and output gaps account for around 75% of all IC failures. Using voltage-based tactics such logic probes, pulsers, and clips is typical for locating problems of this sort. With the use of current tracers, it is possible to efficiently manage repairs to various failure types (see Fig.4.24). A dead driver (open output bond) or lack of pulse activity in the circuit may be the case, for example, if the current through the node is negligible. Use the logic probe and pulser to zero in on the source of the issue. trace the path of the current via the abused node. In many cases, after a defect has been narrowed down to a specific node or group of nodes, the issue is ultimately resolved with the help of a tracer, which is distinguished from the pulser by the fact that it is the last instrument used to isolate the problem.

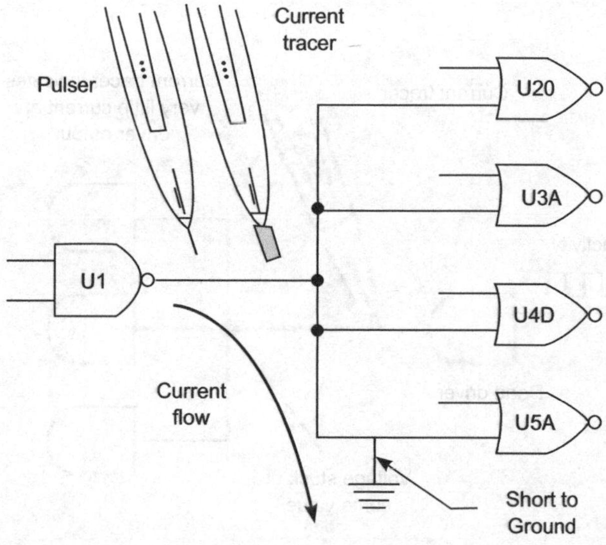

Fig. 4.24: Multiple input fault detection using current tracer and logic pulser

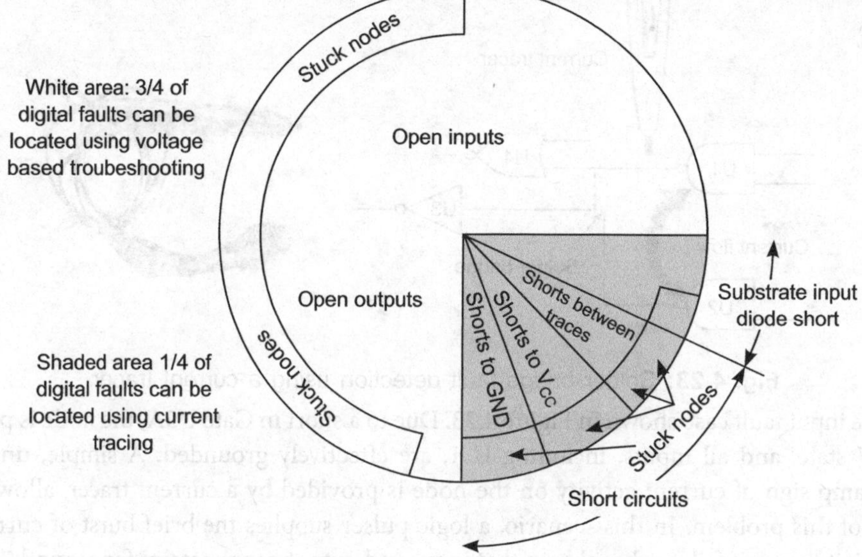

Fig. 4.25: Digital circuit failure modes

Figure 4.25 depicts a common workaround for the issue of a blocked node due to a dead driver. One node has permanently high or low voltage. Space is a must.If the driver is completely dead, or whether some other problem, such a shorted input, is permanently fixing the node's value.This is an easy question to answer with current tracer. The only current shown by the tracer if the driver is dead will be much smaller than and caused by any parasitic coupling from nearby currents.the driver's typical current capacity. However, a typical short circuit will occur if the driver is good. There will be current, which can be followed to the clamping node of the circuit element

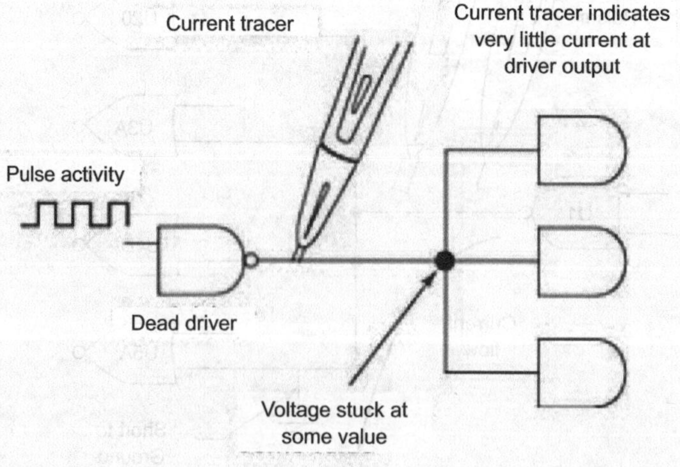

Fig. 4.26: Stuck node caused by dead driver-use of current tracer for fault diagnosis

Vcc-to-Ground Short:

Replacing all of the ICs on the is typically done to try and locate VAC-to-ground shorts either every capacitor on the board or both. Despite most people's experiences, this is a challenging exercise demonstrate that most supply to ground shorts and defective are caused by shorted decoupling capacitors.Most shorts are caused by capacitors. It usually won't pay to, though, if there are numerous bypass capacitors.To find the shorted one and lessen harm to the board, remove each one separately. by means of a these shorts can be quickly identified using the current tracer (Fig. 4.26)

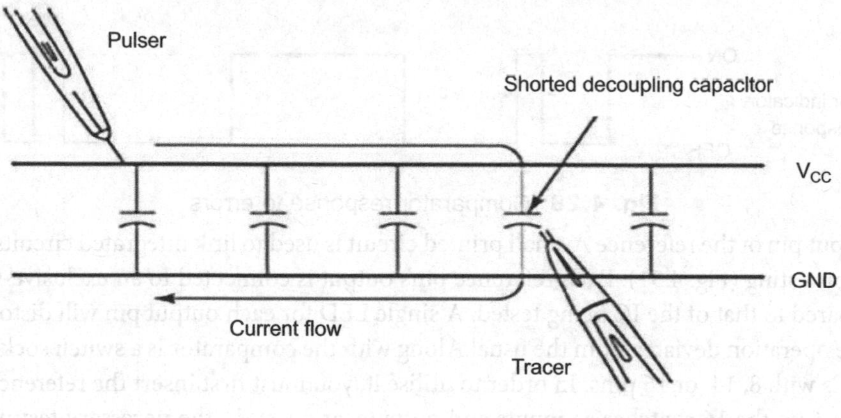

Fig. 4.27: Use of current tracer for VCC-Ground faults

To do this, disconnect the power source and attach the GND lead of a logic pulser to the supply's return terminal. Even if the capacitors are wired in series between the voltage alternating current (VAC) and ground, the current tracer will usually reveal the route along which the most energy is flowing (the shortened path). While voltage-sensing instruments may be used right away, becoming proficient with the current tracer often takes more time. To avoid the cross issue, which happens when tracing a little current in a conductor that is very near to another conductor carrying a much bigger current, takes some expertise. The current tracer's tip sensor may respond to the current in the adjacent trace. However, with some experience,the operator able to identify cross-talk/interference

4.12 LOGIC COMPARATOR

A logic comparator is based on the simple principle that a known-good IC may be used as a yardstick against which the in-circuit performance of suspect ICs can be assessed. TTL/DTL logic problems in the form of malfunctioning nodes or ICs are considerably simpler to locate with this strategy. The comparator does this task and shows any performance issues on a pin-by-pin basis by comparing the output responses of a Reference IC and an in-circuit test IC. Even if the fault is just 200 ns long, the IC output pin that is not correctly following its inputs will still create an indicator that shows error. The comparator indicate mistakes is depicted in Figure 4.27

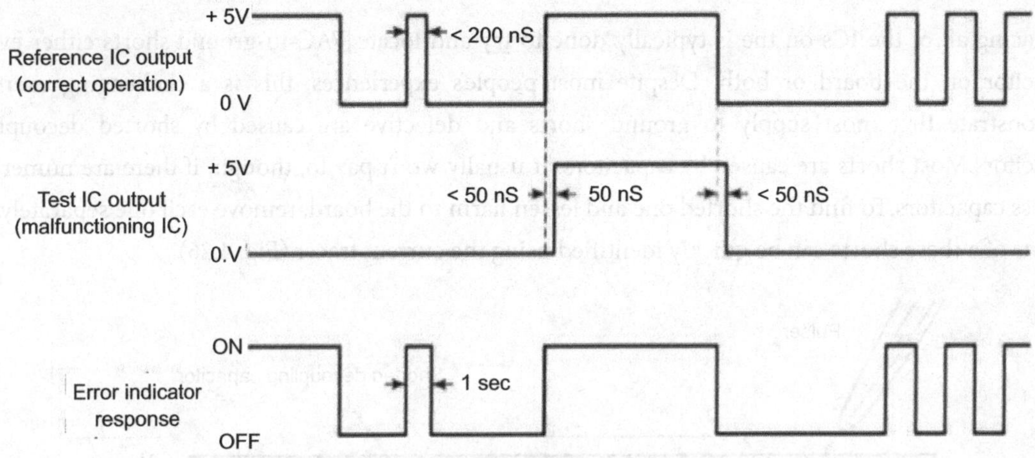

Fig. 4.28: Comparator response to errors

Every input pin of the reference A small printed circuit is used to link integrated circuits in a parallel connection for testing (Fig. 4.31). Each reference pin's output is connected to an exclusive-OR so that it may be compared to that of the IC being tested. A single LED for each output pin will distort and signal an error if the operation deviates from the usual.Along with the comparator is a switch socket board that can handle ICs with 8, 14, or 16 pins. In order to utilise it, you must first insert the reference IC into the socket, then set up the 16 switches as inputs and outputs, and last do the necessary testing. The input pins' supply and ground states are detected by the socket board, and mistakes in the output signal of more than 300 ns are shown on the screen. [1]

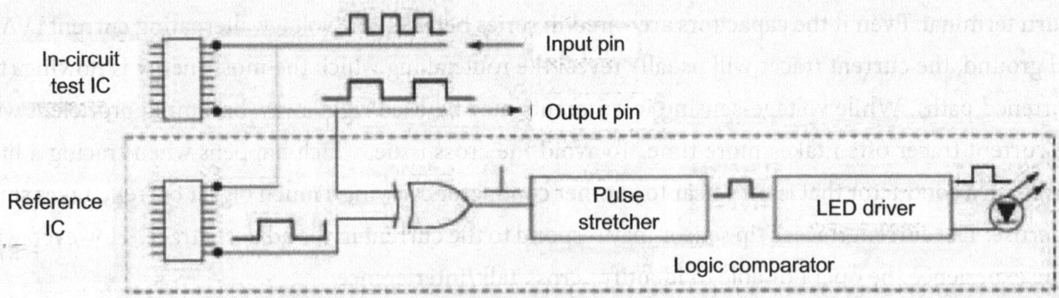

Fig. 4.29: Typical input pin/output pin configuration

The logic comparator's loading effect ensures that it has little to no bearing on the circuit's operation.

However, the use of analogue components like registers, capacitors, or transistors to regulate time or to buffer signals may have an adverse effect on the timing or driving capabilities of these circuits.

4.13 SPECIAL CONSIDERATIONS FOR FAULT DIAGNOSIS

Only if the following conditions are met should digital equipment repair and servicing be attempted.

(a) Pay close attention to the equipment's service manual. Circuits, layout diagrams, part specifications, and a list of the parts should all be included in the manual.

(b) There should be access to the tools and test equipment designed specifically for digital circuits.

(c) Fully comprehend the various logic integrated circuits (ICs) used in the apparatus.

(d) The power supply voltage specifications and expected logic levels need to be known precision.

(e) Power must be on when sending and receiving test signals.

(f) Power supply voltages should be tested at the IC pins themselves, rather than between board connectors on the PCB.

(g) This would prevent the IC from experiencing damage due to a power outage. An IC must not be removed while power is being supplied.

(h) Large test probes should be avoided wherever feasible in digital circuits. These might cause further faults by accidentally shorting off IC pins.

(i) If is no shock hazard, carefully check the temperature of the components with your

(j) When operating, faulty parts usually get hotter than healthy ones.

(k) Use device pins to probe rather than socket pin.

4.14 HANDLING PRECAUTIONS FOR ELECTRONIC DEVICES

1. When testing equipment that are susceptible to static electricity, DC power must be on before, during, and after the application of test signals.
2. When replacing or removing any components or circuit boards, whether they are hardwired or plug in, make sure the power is off to the area.
3. Probably avoid that could cause static charges, such as sitting on a plastic-covered or rubber-footed stool while you comb your hair or using a lot of erasers. These requirements become paramount in dry climates.
4. When keeping or transporting circuit boards or IC packages, be sure to place them in conductive envelopes and/or carriers.
5. Avoid touching the pins while handling IC packages.
6. Use a conductive strap or cable with a wrist cuff to provide a safe and reliable ground connection to the work surface. The cuff must establish complete electrical and physical contact with your skin. It's recommended that sleeves be worn at or below the cuff. The average resistance of skin in contact with a work surface is between 250 kilo volts and one millivolt
7. Avoid using tools and objects that can produce static electricity. Solder suckers with a plunger are an example of such items.
8. The table top, or working surface, must be conductive and safely linked to earth ground with a safety resistance of 250 KΩ to 500 MΩ
9. Connect the tool›s frame directly to the ground if it uses line power. This includes test instruments, lights, drills, soldering irons, and more. Equipment that is grounded should have rubber feet or some other kind of isolation from the work surface. The repairable apparatus must be electrically grounded and electrically insulated.
10. Do not place equipment or electrical components on insulators like books, paper, rubber mats, plastic bags, trays, etc.

4.15 FUNCTION AND TESTING OF FLIP FLOPS COUNTERS AND REGISTERS

4.15.1 Flip-flops

For the most parts, flip-flops will have two distinct states at its output (Fig.4.29 (a)), labeled as Q and it shows true state (Q) and complement (. Signals with either continuous or discrete levels may be fed into the circuit through the input terminals. The most common synonyms for flip-flop are "bi-stable multi-vibrator and latch.

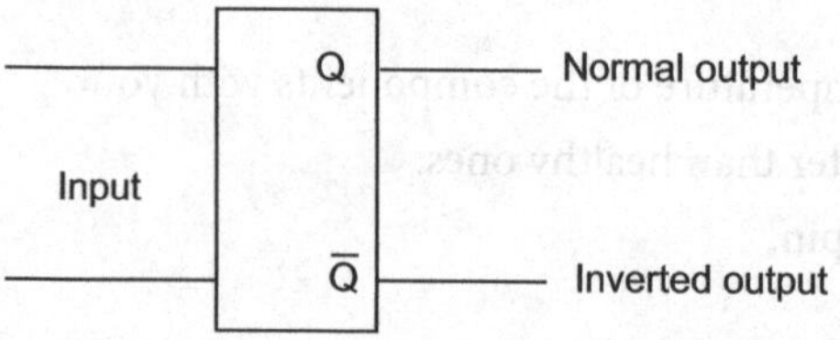

(a)

Outputs		Flip flop
1	0	State
High	Low	1
Low	High	0

(b)

Fig. 4.30: Flip flop configuration and Truth table

The truth table of the flip-flop's showing two potential output conditions and the corresponding definition of the state are shown in Fig.4.29 (b). Flip-flops come in a variety of styles, each with unique attributes. However, the RS flip-flop, a bi-stable component, is present in all those different flip-flop configurations .

4.15.2 Reset-set Flip-flop or the R-S Flip-flop

In (Fig. 4.31), inputs of two NAND gates may be cross-coupled to form a basic flip-flop. Gates A and B produce polarized outputs when given electricity. The Q output from gate A would be 1' if the input to gate B was 1', hence the Q output from gate B would be V. When this 0' is supplied to the input of gate A, the Q output will still read as 1'. As a result, the gates are held in a constant position. Given that Q has a high value, the flip-flop is in its high (1') state. Any extra pulses applied to the SET input have no effect on the output. However, when a pulse is supplied to the RESET input, the output becomes inverted.The outputs are insensitive to any subsequent RESET input pulses. The outputs "flop" back to their original state when the inputs are changed again.

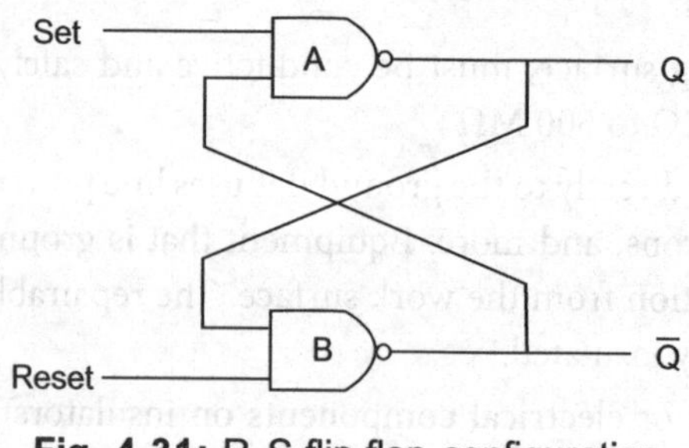

Fig. 4.31: R-S flip flop configuration

The flip-flop functions like a toggle switch, only having one position, and once the switch has been made,repeating the action has no further impact. Regardless, the condition is stable. Although the R-S flip flop has few uses by itself, it is the fundamental building block of flip-flop chains in integrated circuit form. To synchronize the transitions between states, they typically use a clock input. Using a logic probe and comparing the outputs with the truth table, flip-flops are tested for proper operation.

4.15.3 Clocked Flip-flops

The R-S flip-flop is often used when there is no chance of setting and resetting the flop at the same time. If both inputs are activated at the same time, the outputs will go high for the length of a low pulse. The flip flop will change states, perhaps latching in either the 1' or 0' state. The respective time delays of the two NOR gates utilized in the circuit's design determine the stable state that the circuit will finally race to. By using a flip-flop that can only accept input when it coincides with a clock signal, such as one that is synchronous or clocked, the problem of races can be avoided. Applying the correct levels to the inputs enables or conditions the flip-flop, preparing it to transition between states in response to an external pulse. It's the name given to the pulsating signal that serves as a clock and may have originated in an oscillator. An action may be triggered at regular intervals by listening for the clock signal. Figure depicts a sample clock signal (4.31). Operations occur either when the system moves from 1 to 0 (falling edge) or from 0 to 1 (rising edge).

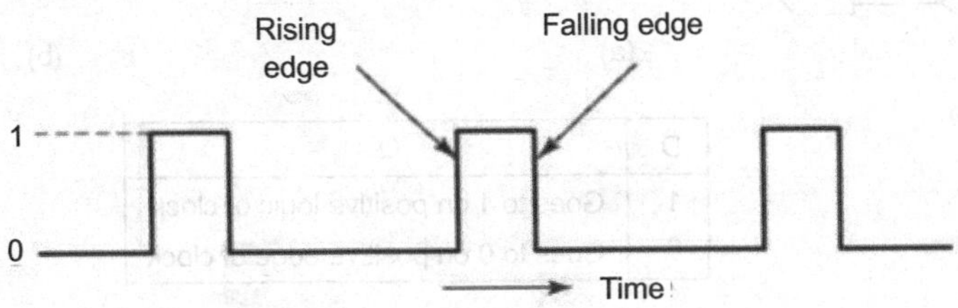

Fig.4.32: Typical clock signal

When the appropriate clock transition occurs, clocked flip-flops are programmed to change states, and they are also meant to idle in between clock pulses. The circuit configuration of a timed R-S flip-flop is shown in Figure (4.35). Gates C and D, which make up the R-S flip-flop circuit, are similar.The SET or RESET input terminal of a flip-flop may also be subjected to a clock pulse in some circuit configurations. If the SET input line is high and the clock input is supplied with a high pulse, gate A will be enabled. If gate A is turned on, the R-S flip flop will be set to the 1' state since it will receive a low SET signal.

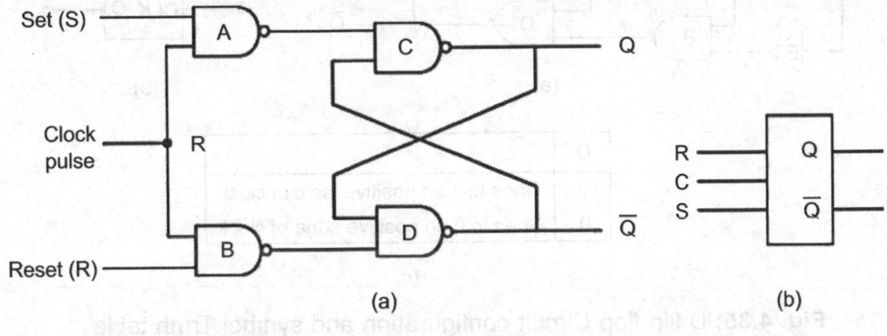

Fig. 4.33: Clocked R-S flip flop symbol and crcuit configuration

If the SET input line is low and the RESET line is active, then Gate B will be activated anytime a high clock pulse occurs. In the 0' state, the flip-flop will be reset by a low RESET signal when gate B is activated. Keep in mind when debugging this circuit that the flip-flop only changes states in response to a clock pulse.

4.15.4 D Flip-flop

Using only one input that can be either high or low is one way to get rid of the intermediate state that is present during the operation of the basic R-S flip-flop. The flip-flop that is produced from this input is referred to as a D flip-flop, or data input. (Fig. 4.33) A D-type flip-flop's circuit layout demonstrates when an input value of 1 or 0 will be transmitted unchanged to the device's input and when an input value of 0 or 1 will be transmitted inverted.

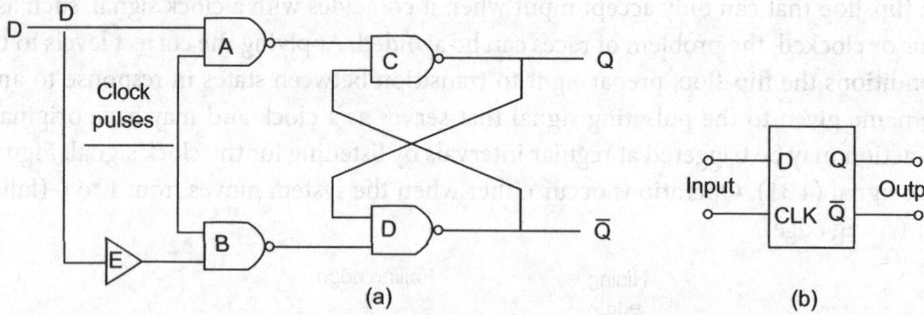

D	Q
1	Goes to 1 on positive logic of clock
0	Goes to 0 on positive edge of clock

(c)

Fig. 4.34: D flip flop Circuit configuration and symbol and Truth table

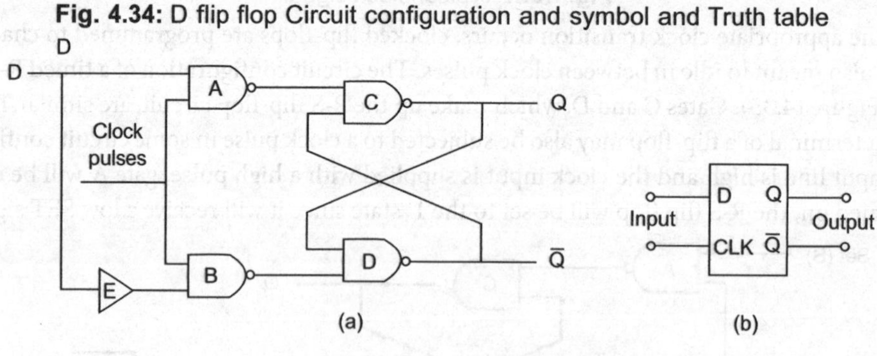

D	Q
1	Goes to 1 on positive logic of clock
0	Goes to 0 on positive edge of clock

(c)

Fig. 4.35: D flip flop Circuit configuration and symbol Truth table

According to Fig. (4.36), when the clock pulse is applied, whatever information was available at the D input before the pulse is sent to the Q output, while its complement is output at the Q output. Therefore, setting the D input to 1' causes the flip-flop to enter the 1' state, while setting it to 0' causes it to enter the 0' state. The edge-triggered D flip-flop is a common form of computer circuit. Like D flip-flops, D-type latches may transition between states while the clock is HIGH ('set), but unlike D flip-flops, when the clock goes LOW (latch), Q retains the state it was in before the transition (and the D input has no effect). As long as there is activity on the CLK line, Q will continue to track the D input regardless of its state.,

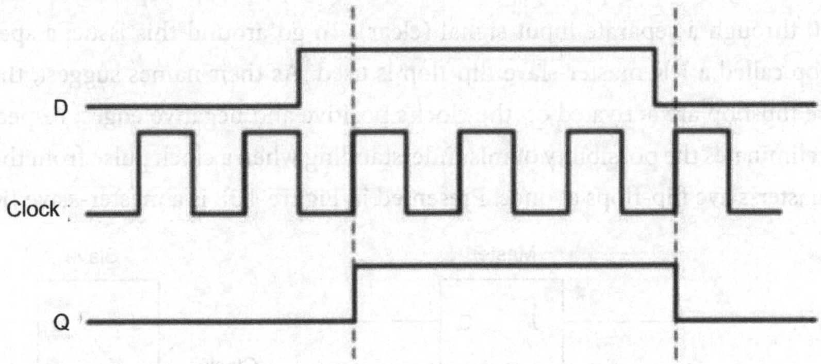

Fig. 4.36: Positive edge triggering in a D flip flop

Analysis of the D-type flip-flop is performed. The flip-flop will not operate without this line being timed. If there is a signal at D and a pulse on the clock line, then there should be a signal at Q..

4.14.5 J-K Flip-flop

The J-K flip-flop is one of the most used logic gates. Unlike other flip-flops, the J-K flip-flop can only be in one of two states. It is the best memory component and the most used form of flip flop in logic circuits. It is common for a J-K flip-flop to have more than one input and sometimes even more than one K input. When used as a clock pulse input, one J input and one K input are commonly connected together. (Fig.4.37.)

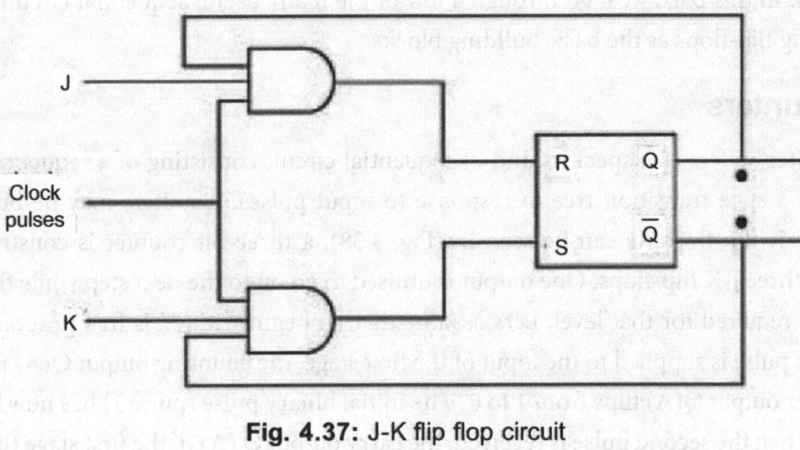

Fig. 4.37: J-K flip flop circuit

- A clock pulse does not change the state of the flip-flop unless both the j and K inputs are active.
- When the flip-J flop's and K inputs are both set, the state of the device is toggled in reaction to the next clock pulse.
- The flip-flop is set using the J and clock pulse inputs, and reset using the K and clock pulse inputs.

The J-K flip-flop is unable to race due to the delay caused by the propagation of information (toggling more than once during a positive clock edge). This happens because of a change in the outputs after a positive clock edge. The output of available commercially J-K flip-flops may be set synchronous to 1 (preset) or 0 through a separate input signal (clear). To go around this issue, a specialized form of the J-K flip-flop called a J-K master-slave flip-flop is used. As their names suggest, the master flip-flop and the slave flip-flop are activated on the clock's positive and negative edges, respectively. This is useful because it eliminates the possibility of misunderstanding when a clock pulse from the same source activates many master-slave flip-flops at once. Presented in Figure 4.38 is a master-slave JK flip-flop.

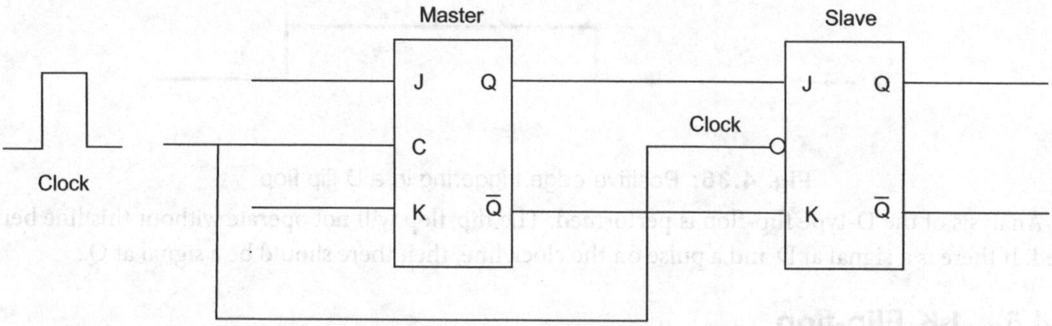

Fig. 4.38: Circuit configuration of master-slave-K flip flop

If the J and C inputs of the J-K flip-flop are coupled together, we get a toggle flip-flop, also known as a T flip-flop, whose output changes state anytime the 1' input is presented. Testing a J-K flip-flop is very similar to examining an R-S circuit, with the addition that the correct signals must be present on the control lines. In this part, we'll go through a few of the many useful sequential circuits that may be constructed using flip-flops as the basic building blocks.

4.15.6 Counters

The word "counter" refers to a specific kind of sequential circuit consisting of a sequence of flip-flops that go through a state transition tree in response to input pulses. Counters may be built out either T flip-flops or J-K flip-flops.As can be seen in (Fig. 4.38), a three-bit counter is constructed from a combination of three J-K flip-flops. One output is utilised to go on to the next step while the other gives the information required for that level. Let's assume that, to begin with, V is free to receive any of Q's outputs. When a pulse is supplied to the input of the first stage, the counting output Q(A) flips from 0 to 1, while the other output Q(A) flips from 1 to 0'. The initial binary pulse (pulse 1) has now been counted by the circuit. When the second pulse is received, the carry output Q (A) of the first stage flips to 1' while the count output QA) returns to 0'. To start counting at 1, the second state (B) is switched. A binary 10 signifies that the circuit has reached the number 2.Stage one is switched to count 1 Q(A) by the third pulse, but stage two is left unaffected. The number 11 in binary notation represents three. This cycle of counting continues until the circuit output is (1,1,1) which is, as we all know, binary notation for 7.

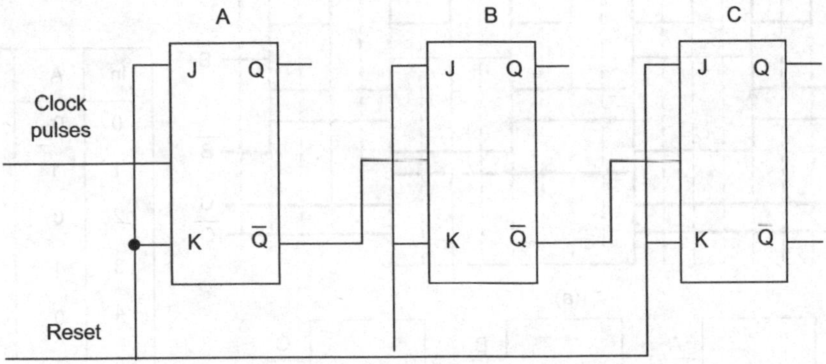

Fig. 4.39: Logic diagram of a 3-bit counter

A circuit reset occurs on the eighth count. Fig. depicts a three-bit binary counter's pulse diagram. 4.39(a).Truth tables for this counter are provided in Figure 4.39(b). When a tenth input pulse is received, a decimal counter, which is a binary counter with four flip-flops and counts up to nine pulses rather than fifteen, resets itself to 0000. A counter of any length can be created by adding additional flip-flops in cascade to the chain. An eight-bit counter is produced by eight flip-flops, a twelve-bit counter by twelve flip-flops, and so on. These counters can only count in increments of one, up to a maximum count, before being reset to zero. Up/down counters are another type of counter that can be designed to count in either direction. The counting direction of these counters is controlled by an up/down input, i.e. The counter counts up from zero when one type of logic level is applied to it, and counts down when the other logic level is applied to it, for example, from 1111 to 0000 (in four-bit counters). Two distinct clock inputs are sometimes used in up/down counters, one for counting upward and the other for counting downward. The commercial marketplace offers a wide range of counters in common integrated circuit packages. They do not require the use of individual flip-flops to build counters. A binary counter and the divide-by- 12 counter DM 7492 are two of the more well-known counters. Synchronous up-down counters are DM 74191, An oscilloscope or a logic probe can be used to quickly test counters and dividers. The truth table is used to identify the activity at each input and output pin, which allows for the isolation of the issue. The output checks from the logic probe should correspond to one line in the truth table.

The outputs are then retested once a pulse has been applied through the pulser. Obviously, the outputs should correspond to the next line in the truth table; if they don't, the chip is defective and has to be replaced. Some counters and divisions need that the "enable" line be brought high or low for the gadget to operate.

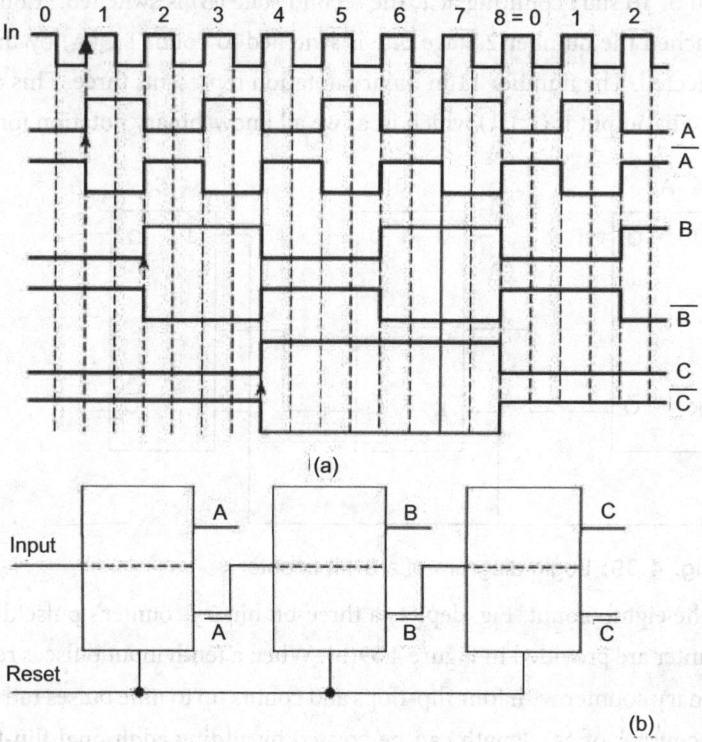

In	A	B	C
0	0	0	0
1	1	0	0
2	0	1	0
3	1	1	0
4	0	0	1
5	1	0	1
6	0	1	1
7	1	1	1
8 = 0	0	0	0

Fig. 4.40: 3-bit counter Timing Truth table and diagram

4.15.7 REGISTERS

Registers are a group of memory components that are utilized when saving binary data. Registers play a crucial role in the transfer of binary information because digital computers rely on binary information being transferred from one register to another and then performing specific operations before it is transferred again. The most fundamental register is a flip-flop. In digital circuitry, parallel latches typically make up a register. The register is controlled by a clock, which tells it when to record input; where n is the total number of parallel latches. The most recent latched data is always shown in the output. The microprocessor's internal registers are essential due to the time-consuming nature of memory access. Intermediate results, for instance, can be temporarily stored in registers rather than being repeatedly returned to main memory. As a result, the number of programmable registers on the CPU is important.

4.15.8 Shift Registers

A shift register is a sequence of flip-flops used to store temporarily. Each clock pulse may also be used to advance the register's contents by one place to the right or left. In order to conduct a left or right shift, the outputs of the flip-flops must be gated to the appropriate inputs. An input mode controls the direction of the shift. It may be more convenient to overwrite the previous entry when you re-enter data into the register. To do this, a specialized "end around shift" function is used. The R-S, J-K, and D kinds of flip-flops may be used to construct a shift register. Connecting several flip-flops together makes the output of one flip-flop the input to the next. A clock pulse shifts the information in the register to the left or right by one place. The TTL or CMOS families of integrated flip-flops are used in their construction, though. In CMOS devices, an integrated shift register can store up to 2048 bits, compared to 4 bits in TTL devices. Only serial-in serial-out CMOS shift-registers are typically used because of the numerous stages needed. There aren't enough pins, so it's impossible to connect pins for parallel input or output

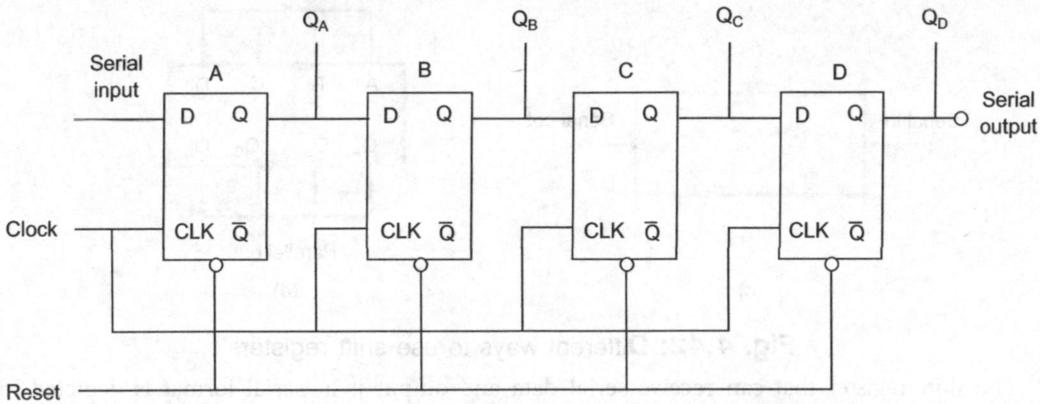

Fig.4.41: 4-bit shift register with serial entry and serial parallel output

An example of a 4 bit shift-register that uses D flipflops is shown in Figure 4.42. What you have here is a shift register with serial input. The outputs at QA, QB, Q, and QD are initially set to 0 by applying a clear pluse means logic 0 to the RESET. Next, serial input is given the first data bit (D_1) to work with. The leading edge of D1 will show a pulse. QA = D_2 and QB = D_1 when the input is updated with the following data bit, D2. After the four clock-pulses QB = Q_3, Q_4 = D_4 ,Q = DZ, and j) = D1 are the results of continuing this process.The circuit described above is a serial-to-parallel converter. In shift registers with a lot of bits, this arrangement is impractical. Serial output registers are used in such situations.

There are four different ways to use shift registers:

1. Output might be serial or parallel.
2. It was either a parallel input and a serial output.
3. A serial input or output
4. Parallel in or parallel out

Data (as shown in Figure 4.43(a)) is supplied serially into a serial in or parallel out shift register, and then, once the entire word has been stored, the bits are studied concurrently from the output of each state. Once all the flip-flop stages have been cleared, the information is sent to a parallel-in/serial-out shift register, as shown in figure 4.43. (b). The information is read out bit by bit in a sequential order at clock

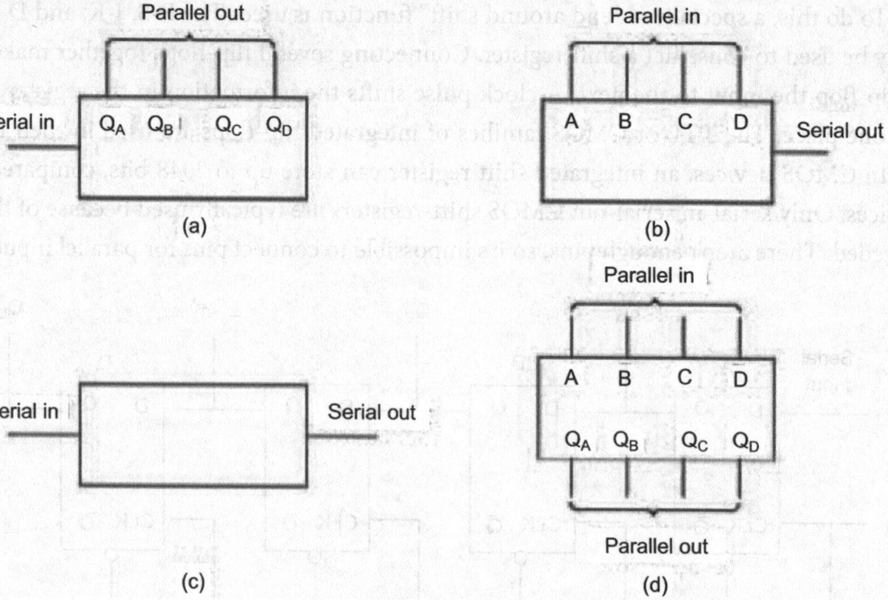

Fig. 4.42: Different ways to use shift registers

The shift register that can receive serial data and output it in serial format is depicted in the figure. In figure 4.43(c), we see a circuit that only acts as a temporary delay. The data is retrieved in the same order as it was saved. Analogous shift registers with parallel inputs and outputs Figure(4.43(d)) illustrates the use of short-term storage. Incorrect control signals are frequently to blame for shift register malfunctions. As an illustration

(a) If the clear line gets blocked, the chip won't accept any new data.
(b) When the clock signal is not there, no change occurs at the outputs.
(c) In order for some shift registers to accept data, an enable signal must be sent to the IC first.
(d) Confirm that this signal is present before attempting to swap out the IC

4.15.9 Multiplexer

A logic circuit called a multiplexer can accept either one data input or one data output, but only one at a time. It basically acts as a switch with multiple positions that can be controlled by the SELECT or ADDRESS inputs. Figure 4.44 provides an illustration of a digital multiplexer.

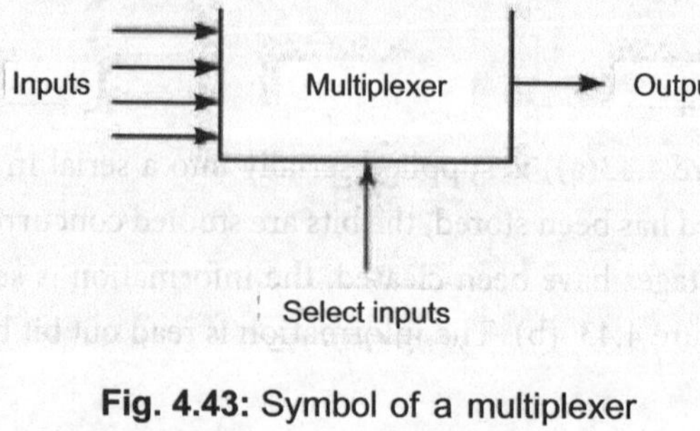

Know more about Multiplexer

Fig. 4.43: Symbol of a multiplexer

Integrated circuit multiplexers are readily available. A multiplexer with 8 inputs and complementary outputs, like the 74151A, is an illustration. Figure 4.45(a) displays the multiplexer's pin layout. The truth table for this multiplexer can be seen in Figure 4.45(b).Other well-liked multiplexers are the dual 4:1 multiplexer 74153 ,16:1 multiplexer 74150.

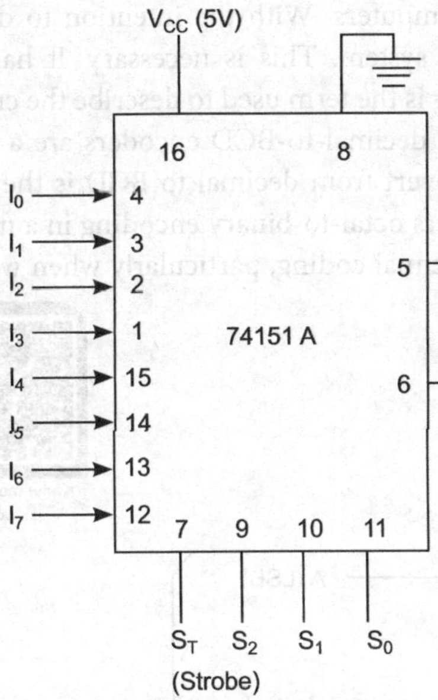

Inputs				Outputs	
Strobe	Select Inputs			y	$\overline{y}$
	S_2	S_1	S_0		
0	0	0	0	I_0	$\overline{I_0}$
0	0	0	1	I_1	$\overline{I_1}$
0	0	1	0	I_2	$\overline{I_2}$
0	0	1	1	I_3	$\overline{I_3}$
0	1	0	0	I_4	$\overline{I_4}$
0	1	0	1	I_5	$\overline{I_5}$
0	1	1	0	I_6	$\overline{I_6}$
0	1	1	1	I_7	$\overline{I_7}$
1	×	×	×	0	1

Fig. 4.44: Truth table and pin configuration of multiplexer (74151A)

4.15.10 De-Multiplexer

The function of a multiplexer is reversed by a de-multiplexer. One input is used, and multiple outputs are created from it. Which output the input data will be sent to is determined by the SELECT input code. Figure 4.46 is a schematic depiction of a de-multiplexer..

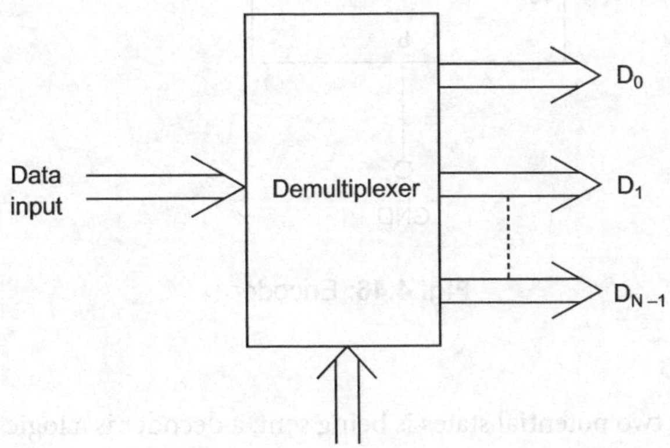

Fig. 4.45: Schematic representation of a de-multiplexer

A dual 1: 4 de-multiplexer, model 74155, is a common illustration of a de-multiplexer. 2-line inputs are changed into 4-line output.

4.15.11 Encoders

While decimal numbers and alphabetic characters are the most widely used forms of communication, binary data is handled in digital circuits, including microcomputers. With the intention to develop interface circuits between the human operators and digital system. This is necessary. It has been developed to convert between different binary codes. Encoding is the term used to describe the creation of binary codes. Hexadecimal-to-binary, octal-to-binary, and decimal-to-BCD encoders are a few of the frequently used codes. The encoder which is used to convert from decimal to BCD is the 74147 IC. It contains the block diagram for 74147. The IC 74148 offers octal-to-binary encoding in a manner similar Figure to 4.47 Microcomputers frequently use hexadecimal coding, particularly when working with lengthy binary words

Know more about Encoder

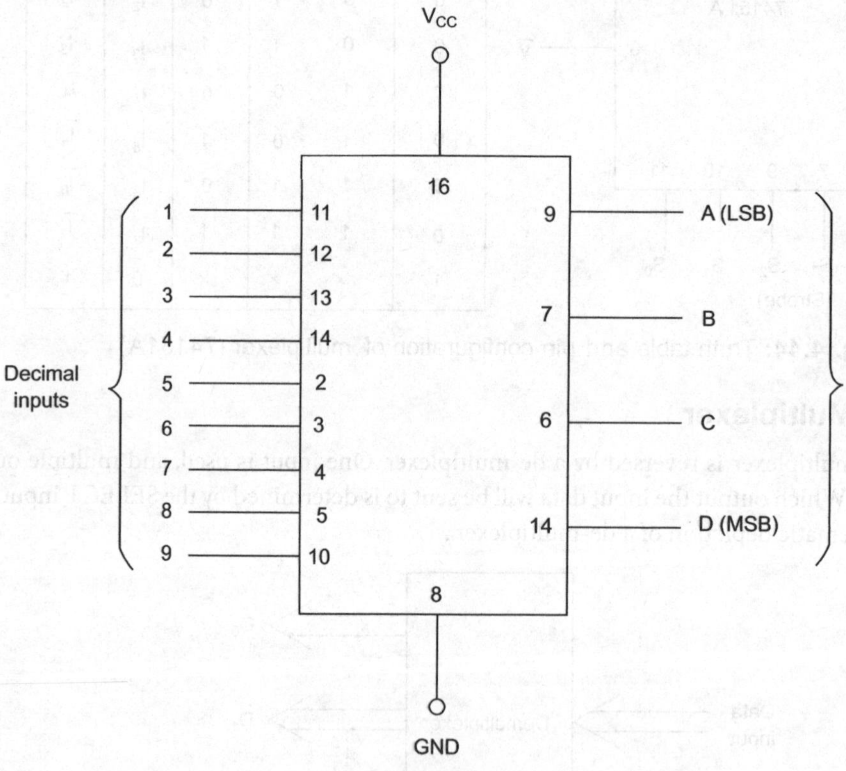

Fig. 4.46: Encoder

4.15.12 Decoders

To determine which of the two potential states is being sent, a decoder is a logic circuit that converts the input 'n' bit binary code into an accurate output signal. The most common decoder is the binary coded decimal to decimal converter, which can translate between four and ten digits per line.Only 10 of the 4-bit binary input's available BCD input codes are used. Ten output pins are present in accordance

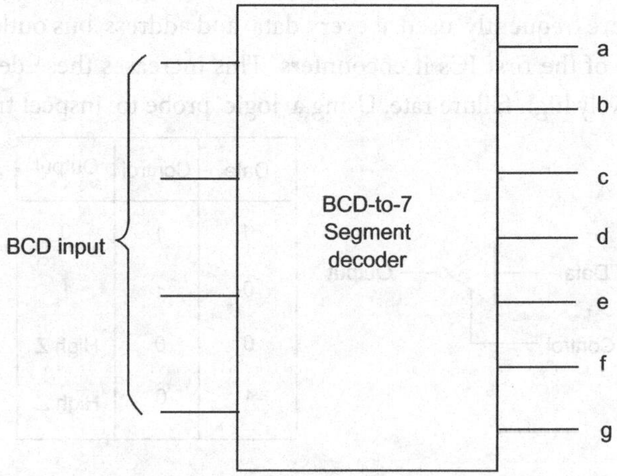

Know more about Tri-state logic

Fig. 4.47: BCD-to-7 segment decoder

In order to convert the BCD code into seven segments, the seven segment LED display unit is often utilized. In view of this, Figure 4.48 shows the diagram of the BCD to seven segment decoder, and Figure 4.49 shows the block diagram of the seven segment LED. This circuit accepts BCD data on four input lines and outputs it as seven segment display lines. To properly configure a display, the decoder's a-through-g outputs must be linked to the corresponding a-through-g inputs. It is possible to have either active low or active high outputs from the decoder, and the display's seven segments may be either seven cathodes with a common anode or seven anodes with a common cathode

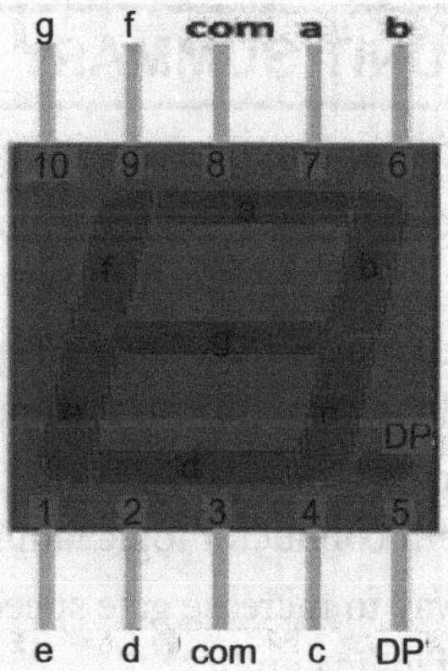

Fig. 4.48: LED display

4.15.13 Tri-state Logic

As its title suggests, tri-state gates can only produce one of three possible states as an output. In addition to the conventional high (1 for 1) and low (0 for 0) logic states, tri-state gates provide a third state with very high output impedance. Fig.4.50 shows the range of possible configurations for tri-state buffers. No data is being processed by the apparatus, and the control signal is at an all-time high. When the control line is in its low state, the switch is "on," and the input signals go to the output. In microprocessor-based circuits, tri-state buffers are frequently used at every data and address bus outlet. When a signal enters the computer, this is one of the first ICs it encounters. This increases these devices' vulnerability and contributes to their relatively high failure rate. Using a logic probe to inspect tri-state buffers.

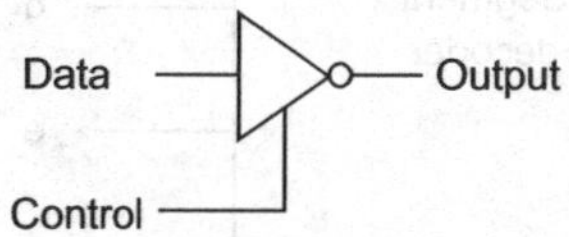

(a)

Data	Control	Output
1	1	0
0	1	1
0	0	High Z
1	0	High Z

(b)

Fig. 4.49: TTL tri-state invertor symbol and truth table

Check following:

(a) The input and output pins should be displaying the same signal if the tri-state buffer is enabled.

(b) Despite having a good input, the output from this inefficient buffer is also nil. An unhealthy buffer also causes intermittent high or low output. Typically, eight tristate buffers will be placed on a single chip. Whenever even a single stage of an integrated circuit (IC) fails, the whole IC must be changed..

UNIT SUMMARY

- Digital IC's are mostly used in every areas of electrical or electronics feilds, including automation industrial control, electronics circuit design instrumentation and communication systems, or medical equipment manufacturing.
- With the development of integrated circuit technology, the problem of size has been overcome, and it is now possible to acquire several functionalities on a chip.
- Most often seen and used in modern digital devices are circuits from the transistor-transistor logic and complementary metal-oxide semiconductor logic families.
- The ECL family offers a different way to increase gate speed. The transistors make this completely different from the other varieties of logic families.
- In order to determine which of the possible two combinations is present, a decoder change the input binary code into the exact output signal.
- Tristate gates are created to provide output in different three states. In addition to the typical logic 1 high and logic 0 low states, tri-state gates also provide a third state with a very high output impedance.
- A de-multiplexer performs a multiplexer's opposite function. It splits an input into several outputs after receiving a single input.
- Encoding is the term used to describe the process of creating binary codes. Hexadecimal-to-binary, octal-to-binary, and decimal-to-BCD encoders are a few of the frequently used codes.

- To briefly store data, a group of flip-flops may be linked in serial to create a shift register.
- The main purpose of a logic current tracer is to detect current sources or sinks in order to locate low-impedance faults in digital circuits.
- The JK flip-flop is an exceptionally useful member of the flip-flop family. The J-K flip-flop stands out since it can only be in one of two or more states.
- Internal transistor open circuits or internal open connection are two potential faults. The OV line is considered to be externally sound.With a logic probe or an oscilloscope, counters and dividers can be quickly checked.
- The term "integrated circuit packaging" describes how a semiconductor component is enclosed. Any IC's central component is made up of semiconductor wafers that are layered with copper and other materials in a sophisticated pattern.

EXERCISE

1. What is the name of the logic family that emits the least amount of power?
   (a) TTL (b) CMOS
   (c) ECL (d) DTL
2. What is the name of the logic family with the lowest fan-out value?
   (a) Standard TTL (b) CMOS
   (c) ECL (d) Schottky TTL
3. What is name of that logic family who have the highest fan-out?
   (a) CMOS (b) TTL
   (c) Schottky TTL (d) ECL
4. Which statement is not true .
   (a) The J-K flip-flop is has no ambiguous state.
   (b) Basic gates, NAND & NOR are universal gates.
   (c) The inputs for logic gates will be an analog signal
   (d) Logic family are useful in very high-frequency applications
5. The logical gates are categorized into which group
   (a) One (b) Two
   (c) Three (d) Four
6. Following is the advantage of static complementary gates?
   (a) Reliable (b) Not easy to use
   (c) Not reliable (d) Reliable and easy to use
7. Who discover Boolean algebra
   (a) Scientist Bardeen (b) Scientist Claude Shannon
   (c) Scientist George Boole (d) None of the above
8. What is identity law
   (a) a + 0 = 0 + a = a (b) a + 1 = 1 + a = 1
   (c) ab = ba (d) None of above

9. What is the 7400 TTL family' maximum toggle speed
   (a) 20MHz (b) 10 MHz
   (c) 15 MHz (d) 25 MHz
10. What is the speed of the low power Schottky
   (a) 2ns (b) 8ns
   (c) 10ns (d) 12ns
11. Which is the instrument who often used to detect logic levels
   (a) Logic meter (b) Logic pulser
   (c) Logic probe (d) logic comparator
12. What is the name of combinational logic circuits which has 2^n input line ans single output line
   (a) Multiplexer (b) Demultiplexer
   (c) Decoder (d) Encoder
13. What is not showing output state of tri state logic
   (a) High (b) Low
   (c) High z (d) Low z
14. If inputs are 1,1 then what will be output of J-K flip-flop
   (a) 1,0 (b) 0,1
   (c) 1,1 (d) All above
15. During the manufacturing of NOR and NAND gates the Latches wish to stay within the latched condition ,which is following feature is accountable
   (a) Low voltages as input (b) Synchronous Operation
   (c) Gate Impedance (d) Cross coupling

ANSWERS															
1.	(b)	2.	(a)	3.	(a)	4.	(c)	5.	(c)	6.	(d)	7.	(c)	8.	(a)
9.	(d)	10.	(c)	11.	(c)	12.	(c)	13.	(d)	14.	(a)	15.	(d)		

Short and Long Answer Type Question

Category 1

1. Describe the voltage levels of Digital signal in both positive and negative logic.
2. Explain applications of Logic Probe.
3. What is meaning by Digital integrated circuits?
4. Which device can be used as identifier,and explain why?
5. Define logic comparator.
6. Describe mostly appeared faults in digital circuits.
7. How to detect two mutually shorted integrated circuits pins?
8. Explain the difference between logic pulser and logic probe.
9. What is advantage of logic clip.
10. What are the procedure of identification a digital IC from its number?

Category 2

1. Explain the various packages in which the digital ICs are available.
2. Describe various faults in digital circuits.
3. Study about combinational faults with their example.
4. What are the basic precautions required during digital troubleshooting process
5. Explain the working procedure and construction details of following
   (i) Logic comparator (ii) Logic pulser
   (iii) Logic probe (iv) Encoder

REFERENCES AND SUGGESTED READINGS

1. Khandpur, R.S. (1996), troubleshooting of instruments Tata McGraw- Hill Publishing Co. Ltd., New Delhi. ISBN 0-07-460186-5
2. Digital Design by M.Morris Mano
3. Digital Circuits and Design by Salivahanan S.
4. Handbook of Components for Electronics, McGraw-Hill Co., USA, ISBN 0-07-026682-4.
5. Integrated Circuits And Logic Families by Jack Kilby in 1958 O'Reilly learning platform.
6. Digital Logic Circuit Families,ByNassir H. Sabah,BookElectronics,Edition1st Edition,First Published2010,imprint
7. Troubleshooting Electronic Circuits: A Guide to Learning Analog Electronics: by Ronald Quan McGraw-Hill, U.S.A.
8. Loveday, G (1989), Electronic Testing and Fault Diagnosis, Longman (Scientific and Technical), England. ISBN -582-03865-0
9. Malvino, A.P. (1989), Electronic Principles, 4th ed. McGraw-Hill, U.S.A. ISBN 0-07099479-X.
10. Scarlett, J.A. (1984), An Introduction to Printed Circuit Board Technology, Electrochemical Publications Limited. U.K. ISBN 0-901150-16-9.
11. http://nitttrc.edu.in/nptel/courses/video/106105185/L51.html
12. http://www.gtu.ac.in/Syllabus/New_Diploma/sem-2/Pdf/3322402.pdf
13. https://www.youtube.com/watch?v=pI0pLx6oPd8

KNOW MORE

- The ordinary longevity of seven-segment display case differed from 5, 000 hours to 200, 000 hours .
- In telephone network, multiple audio signals are integrated on a single line for transmission with the help of multiplexers.
- The transistor, is a mixture of the words transfer and resistor

QR SCAN CODE FOR SUPPORTIVE KNOWLEDGE EMBEDDED IN THE CHAPTER

5 Repairing of Surface Mount Assemblies

UNIT SPECIFICS

The following topics have been covered in this unit:

- Surface Mount Assemble Rework and Repair
- Devices and Surface Mount Technology.
- Semiconductor Surface-Mount IC Packages
- The Flat Pack and the Quad Pack
- Diode Packages in a Cylindrical Shape
- Problems related to repairing Surface Mount PCBs and their rework stations.

The unit has a significant number of multiple choice questions, as well as questions with short and long responses, grouped into two categories that correlate to the lower and higher orders of Bloom's taxonomy, to promote more investigation, creativity, and problem-solving skills. It also includes a list of references and recommended readings that discuss the real-world applications of these issues. It's worth noting that QR codes, which may be scanned for further information, have been included in various parts so that readers can learn more about the themes that interest them.

Related exercises are followed by a "Know More" section that expands on the material covered in the exercises. This appendix has been thoughtfully constructed to ensure that the additional material included here is useful to readers of this book. This part focuses on the topic's origins, provides instances of relevant facts and analogies, traces its historical development while emphasizing key observations and findings, provides timelines that span the evolution of the relevant subjects from their earliest stages all the way up to the present day, and discusses the topic in terms of its relevance to our everyday lives and the ways in which it may be used in the workplace..

RATIONALE

This unit on surface mount assemblies helps students to get a general knowledge about the maintenance of surface mount device and their faults repairing procedure. It explains the concept behind cylindrical diode packages, as well as flat packs and quad packs. The purpose of Rework and Repair of Surface Mount Assemble is to keep them in working condition.

PRE-REQUISITES

Physics: Electronic devices ,Basic knowledge about semiconductors

UNIT OUTCOMES

List of outcomes of this unit is as follows:

U5-O1: Discuss the rework and repair of Surface Mount Assemble

U5-O2: Describe the Surface Mount Technology and devices.and their packages

U5-O3: Explain Flat packs and Quad Packs

U5-O4: Demonstrate Cylindrical Diode Packages

U5-O5: Interpret Problems related to repairing Surface Mount PCBs and their Rework Stations

Unit-5 Outcomes	EXPECTED MAPPING WITH COURSE OUTCOMES (1 - Weak Correlation; 2 - Medium Correlation; 3 - Strong Correlation)					
	CO-1	CO-2	CO-3	CO-4	CO-5	CO-6
U5-O1	1	1	1	2	2	1
U5-O2	2	2	2	3	2	2
U5-O3	3	3	3	2	3	1
U5-O4	2	1	2	3	1	2
U5-O5	3	3	3	-	2	2

5.1 INTRODUCTION

Surface mount technology is rapidly replacing through-hole technology used in PCB fabrication.It is especially easy to automate the assembly of the surface-mounted parts because of their packaging. Surface mounting's benefits include streamlined manufacturing, smaller circuit boards, and higher dependability. The ever- increasing need for miniature electronic assembly, especially portable items, is the primary impetus for using surface mount technology (SMT). Among this technology's benefits are:

- Reduced Board Size
- Smaller PCBs are needed for smaller products.
- Parts that may be attached to either side.
- Great Product:
- More capabilities in a smaller package
- Mechanical and electrical efficiency have been increased.
- The Price of Production Is Reduced
- The cost of SMT parts is lower;
- Reduced number of parts due to the space-saving nature of multi-function ICs;
- Machines used in production are standard, and they are highly automated.
- The costs of materials, labour can be reduced in half or more.
- There is less potential for component damage or waste because to streamlined component handling.
- EMI/EMC issues will be further reduced by using surface-mount devices (SMD) instead of leaded devices.

Know more about SMD

Fig. 5.1: Surface mount devices

5.2 SURFACE MOUNT DEVICES

Surface mount devices, sometimes abbreviated as "SMD,". he most prevalent kind of component utilized in this technology. Smaller and more compact than their leaded counterparts, SMDs include soldering pads or short leads for easy assembly. SMDs, or surface-mount gadgets, are components that are connected to the PCB's surface without the need of any holes. Around 80% of all SMDs are discrete semiconductors, ceramic capacitors, inductors and resistors. As cylindrical SMDs may only have two pins, they are often primarily used for resistors, capacitors, and diodes, giving rise to the cubic shape's predominance in the SMDs. It is possible to bend the leads of a DIP package to make it surface mountable if building a dedicated SMD package is not feasible for technical or economic reasons. Directional Methods SMDs may have a variety of lead styles because to their construction. Figure 5.2 displays the most popular types of lead .

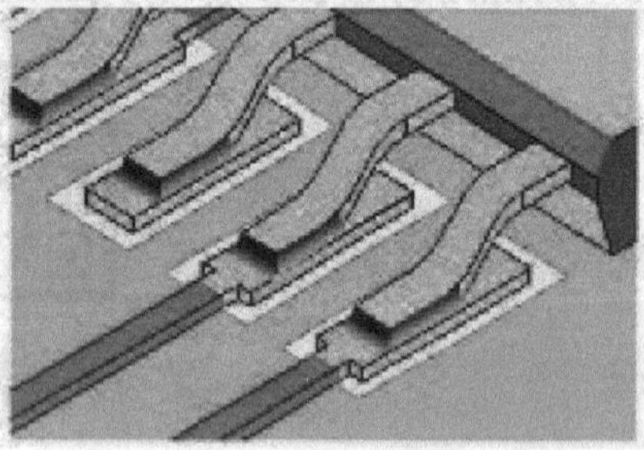

Gull-wing lead

J type lead

Fig. 5.2: SMD lead types

Lead Pitch

As can be observed in Fig. 5.3, the lead pitch of an SMD is really the distance between the lead centres rather than the air gap between them.

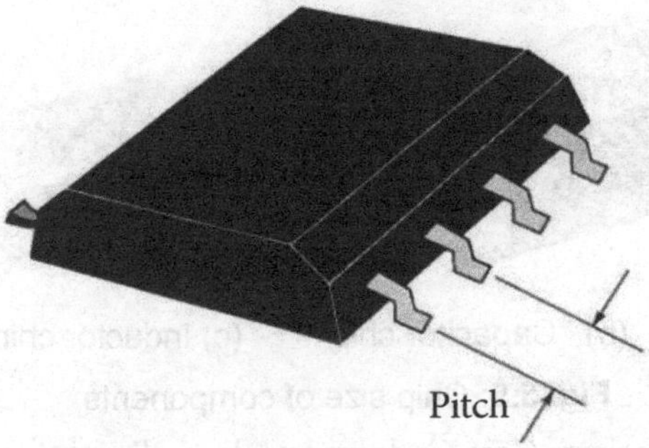

Fig. 5.3: SMD packaging techniques

Component package technique

The use of pick and place machines has greatly aided the automated assembly of printed circuit boards. To this end, the components, especially the SMD s, must be properly packaged to prevent damage during transit. Bulk feed cassettes, tape and reel, and trays and tubes are only some of the packaging options for components. As depicted in Fig.5.4.

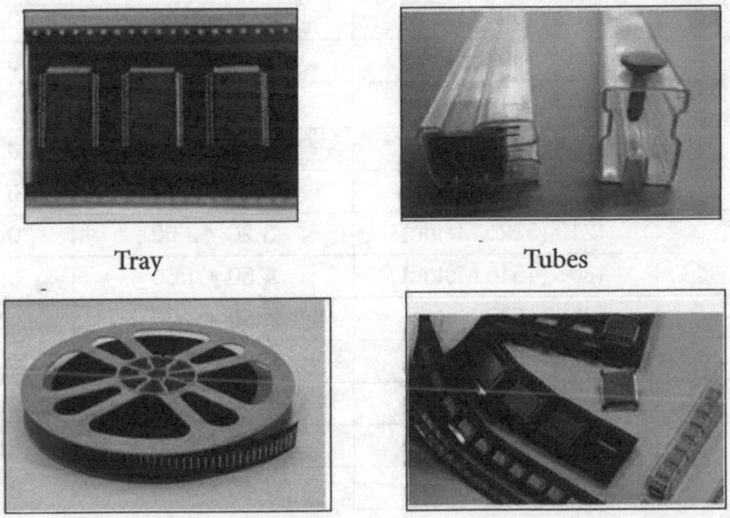

Tray　　Tubes

Tape and Reel　　Bulk feed Cassettes

Fig. 5.4: SMD packaging techniques

Chip Size Codes

A 4-digit size code specifies the dimensions of a chip's individual components. The size code does not provide the thickness of the component. Component size specification examples are shown in (Fig.5.5.) The size designation may be given in either the imperial or metric system. Ceramic capacitors and resistors, for instance, typically list their size in inches, whereas tantalum capacitors list their size in millimetres.

(a) Resistor chip (b) Capacitor chip (c) Inductor chip (d) Diode chip

Fig. 5.5: Chip size of components

Here is a table listing the Chip package Case Code and package dimensions for "Standard Chips"

Table 5.1: Chip package dimensions

Inch Case Code	Inch (Metric) Case Code	Metric Dimensions	Inch Dimensions
008004	008004 (0201 Metric)	0.25 × 0.125	0.010 × 0.005
009005	009005 (03015 Metric)	0.30 × 0.15	0.012 × 0.006
01005	01005 (0402 Metric)	0.40 × 0.20	0.016 × 0.008
0201	0201 (0603 Metric)	0.60 × 0.30	0.020 × 0.010
0402	0402 (1005 Metric)	1.00 × 0.50	0.040 × 0.020
0603	0603 (1608 Metric)	1.60 × 0.80	0.060 × 0.030
0805	0805 (2012 Metric)	2.00 × 1.25	0.080 × 0.050
1008	1008 (2520 Metric)	2.50 × 2.00	0.100 × 0.050
1206	1206 (3216 Metric)	3.20 × 1.60	0.125 × 0.060
1210	1210 (3225 Metric)	3.20 × 2.50	0.125 × 0.100
1806	1806 (4516 Metric)	4.50 × 1.60	0.180 × 0.060
1812	1812 (4532 Metric)	4.50 × 3.20	0.180 × 125
1825	1825 (4564 Metric)	4.50 × 6.40	0.180 × 0.250
2010	2010 (5025 Metric)	5.00 × 2.50	0.200 × 0.100
2512	2512 (6332 Metric)	6.30 × 3.20	0.250 × 0.125
2920	2920 (7451 Metric)	7.40 × 5.10	0.290 × 0.200

5.3 SURFACE MOUNTING SEMICONDUCTOR PACKAGES

Different varieties of packaging have arisen with the evolution of surface-mounting technology. Common semiconductor Surface Mount Device packages include the following:

5.3.1 SOIC (Small Outline Integrated Circuit)

This plastic casing comes in a variety of pin counts and widths, including 6, 8, 10, 14, and 16 pins with a body width of 7.6 mm and 6, 8, 10, 14, and 16 pins and a body width of 4 mm. All 1.27 mm of the lead ends make contact with the PCB (Fig. 5.6). The two additional very small outline (VSO) packages feature either 40 or 56 leads spaced at 0.762 mm.

Fig.5.6: Small Outline Integrated Circuits (SOIL)

5.3.2 SOT (Small Outline Transistor)

Transistors and diodes that are not part of a circuit board often come in these packages. SOT-23 and SOT-89 are the two most common packages (now renamed as TO-236 and TO-243 respectively). Standard SOT-23 Specifications and Dimensions are Detailed in (Fig.5.7). The package's three leads are located as follows: twice along one edge and once in the middle of the other. SOT-89 packaging is used for semiconductors with bigger chip sizes (up to around 1.5 mm square) (Fig.5.8). The central lead crosses the base to enhance heat conduction, while the other two leads are arranged in a line down one side of the package.

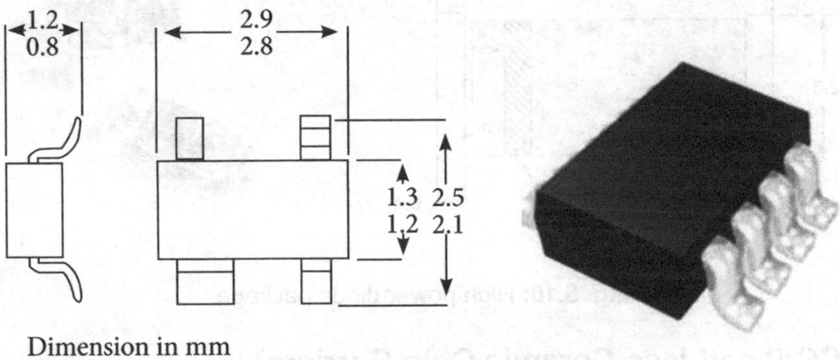

Fig. 5.7: SOT package(SOT-23)

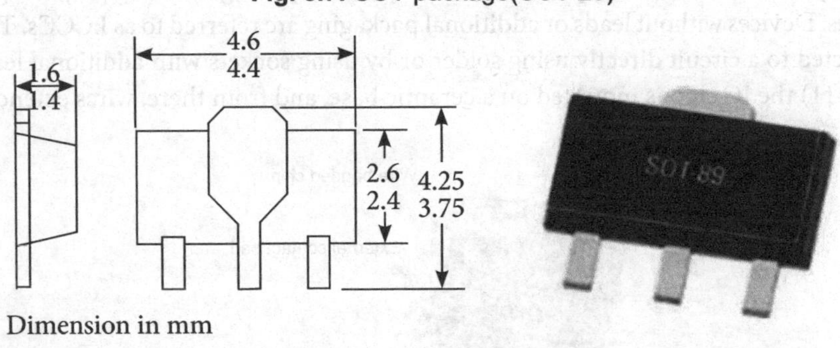

Fig. 5.8: SOT-89 package

5.3.3 Cylindrical Diode Packages

These are the two most common packaging, both of which were specially designed for cylindrical diodes.

1. The SOD (Small Outline Diode) packaging was developed for diode chips with a maximum power dissipation of 250 mW. The dimensions of a representative device of this kind, the SOD-80, are seen in (Fig. 5.9.)

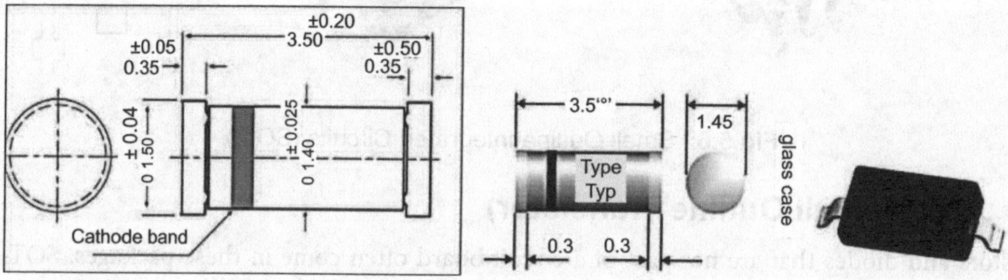

Fig. 5.9: Small outline diode (SOD-80)

In cases when more power dissipation is required, a MELF (Metal Electrode Face Bonded) package with a more robust cylindrical casing is employed. As may be seen in (Fig. 5.10) MELF diodes typically have the size described there. Mini MELF is another name for the SOD-80 package.

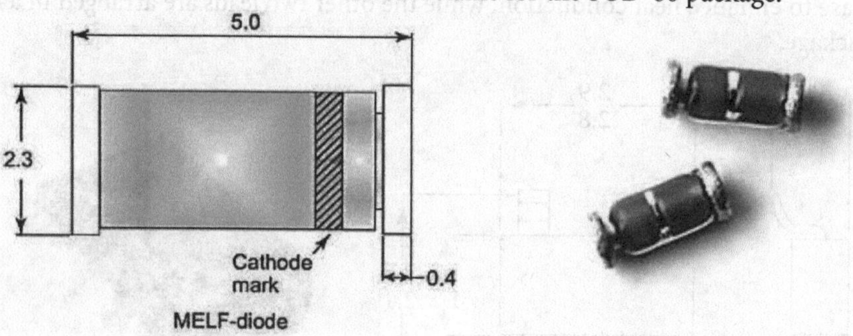

Fig. 5.10: High power diode package

5.3.4 LCCC (Lead-less Ceramic Chip Carriers)

The phrase "chip carriers" refers to square or nearly square IC packages having projecting terminations on all four sides. Devices without leads or additional packaging are referred to as LCCCs. These packages may be connected to a circuit directly using solder or by using sockets with additional leads. As can be seen in(Fig. 5.11) the IC chip is mounted on a ceramic base, and from there, wires extend to solderable contact pads.

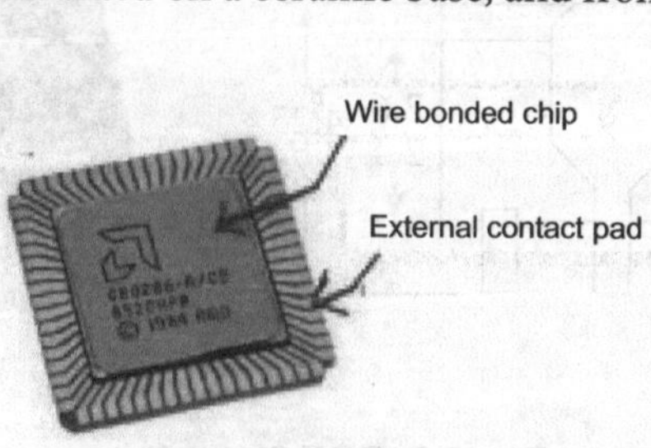

Fig. 5.11: Lead-less (LCCC) ceramic chip carrier

The most frequent termination lengths for LCCCs are 18, 20, 28, 32, 44, 68, 84, 100, 124, and 156. Height of componentsis normally between 1.5 and 2.0 mm. Constantly 1.27 mm is the distance between the ends.

5.3.5 Plastic Leaded (PLCC) Chip Carriers

They come in a broad variety of sizes and configurations, paralleling those of LCCCs. The devices' leads likewise feature a 1.27 mm pitch. J' leads, which are folded beneath the package, are standard on PLCCs. Observe this in (Fig.5.120 'J' leads cause problems during circuit inspection and testing since they are not protruding from the device.

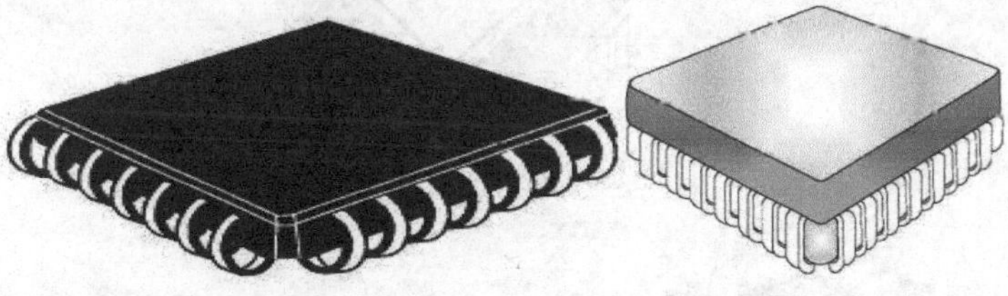

Fig. 5.12: f-leaded PLCC package

5.3.6 Flat Packs and Quad Packs

The lead frame of the flat pack runs parallel to the flat package.As opposed to the original flatpack, which only featured leads on two sides, the newer version, known as a "quad pack," has them on all four sides. Typically, the lead count of the plastic containers used for quad packs is in the region of 40 to 200. The number of leads has no effect on the overall size of the package, but the distance between them does. For smaller packages (up to 64 leads), the pitch is 1.0 mm, for larger packages (80 leads and 100 pins), it is 0.8 mm, and for the largest packages (100 pins), it is 0.65 mm. As may be seen in Fig. 5.13, the package has the form given there.

Fig. 5.13: Quad flat pack package

5.3.7 LGA (Land Grid Arrays)

Instead of being located along the device's sides, the pins in this type of device radiate from a grid on the device's underside (Fig. 5.14). Land grid arrays are a type of surface-mount IC package that eliminates the need for pins by replacing them with solderable pads. To support a broad range of lead-out configurations, they are available in a variety of forms, pad sizes, and pad densities.

Fig. 5.14: Land Grid Array

5.3.8 BGA (Ball Grid Arrays)

The pitch required to connect the peripheral I/Os can be a problem for some surface mount components, including quad flat packs (QFPs). To accommodate additional input/outputs, the package pins must have a finer pitch. The component lead pitches have been lowered from 0.5mm to 0.3mm, and the number of available I/Os has been increased. Because of how easily the leads can be broken, this component is at risk of damage during transit or PCB assembly. Integration of these components into PCBs also necessitates control. The challenges of using fine pitch peripheral leads have been mitigated by the advent of the ball grid array (BGA), which allows for the fabrication of area array interconnects. By employing a solder ball array in place of pins, BGA technology allows for high-density connection options for packages with 250-1,089 I/Os in the same volume as a 208-pin quad flat device (QFP). However, while the shift to 1.27mm pitch has fixed the issues with 0.3mm QFPs, it has also introduced new issues. Multi-level substrates are now required to enable the I/Os to fan out from the device, and more precise inspection methods are required to look at the buried solder connection. However, BGA assemblies are rapidly replacing other packaging methods for high I/O devices that require more reliable electrical connections. Each of the four primary categories of BGA packages may be identified by the array of solder balls that is affixed to the base of the package body or carrier [2]. There are a wide variety of possible configurations for this packaging method, including the "Slightly Larger than IC Carrier" (SLICC) package and the micro ball grid array (BGA).

- **Solder ball attach**

The solder balls in a ball grid array package may be joined to the carrier in a few different ways. Two basic approaches involve printing solder paste and then reflowing it into a ball, or putting a preformed solid core solder ball into soft solder already present on the carrier

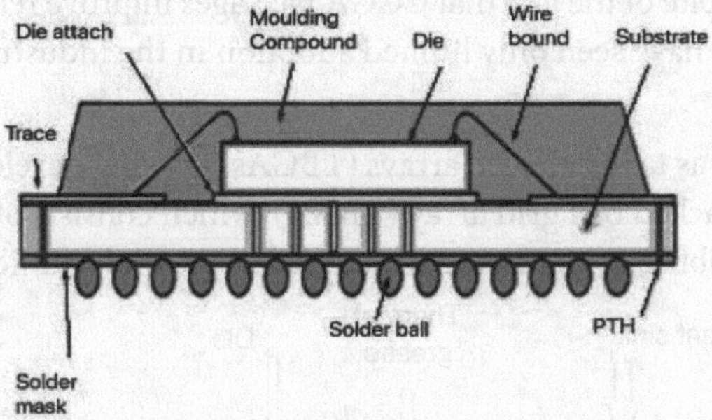

Fig. 5.15: Plastic bag grid array

- **Plastic ball grid arrays**

Most BGA packages, as seen in (Fig.5.15) are plastic ball grid arrays (PBGAs), also known as OMPACs (Over Moulded Plastic Array Carriers) The PBGA carrier is often made from FR4 or BT resin, two types of printed circuit board material. After the die has been wire-bonded to the top surface of the PCB carrier, a plastic over-mold is often added (epoxy based). In plastic devices, the PCB carrier's bottom is typically soldered with solder balls of between 0.5 and 0.7 mm in diameter. Tin-lead eutectic or tin-lead-silver alloy is used in their production. Balls can also be made by printing .

- **Ceramic ball grid arrays**

Another type of BGA packaging is the Ceramic Ball Grid Array (CBGA), which is also known as a Solder Ball Carrier. CBGAs, depicted in(Fig. 5.16,) involve attaching a die to the upper surface of a ceramic multi layer carrier. Wire bonding the die with the active side up is one option, while flip-chip attachment allows the die to be connected with the active side down. When the die is attached, it is encased to increase its durability and shield it from actual injury. High-temperature (solid core) solder balls are melted when they are reflown onto the carrier of a ceramic package using eutectic or close-eutectic tin-lead or tin-lead-silver paste (90Pb10Sn).In terms of dimension, the solder balls are around 0.8 millimeter in size. The high- temperature ball remains intact throughout the reflow, providing a consistent clearance distance (controlled collapse connection).

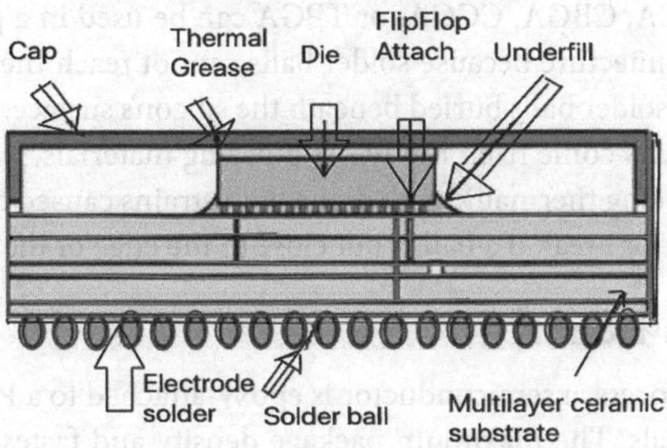

Fig. 5.16: Ceramic ball grid array

Know more about Rework station

- **Ceramic column grid arrays**

CBGAs have an alternative in Ceramic Column Grid Arrays (CCGAs) for ceramic bodies larger than 32mm 2 . The ceramic carrier has a series of columns of high temperature 90Pb10Sn solder affixed to the bottom of it rather than solder balls. The diameter of the columns is typically 0.5 mm, while their height is typically 1.75 mm. Under spite of the fact that CCGA packages improve reliability and stand-off height in heat cycle conditions, they have seen only limited adoption in the industrial sector.

- **Tape ball grid arrays**

New BGA packaging, known as tape ball grid arrays (TBGAs) is being developed by companies like 3M. In (Fig. 5.17), we can see a stacked ball grid array (TBGA), which consists of a die mounted on a carrier (often a copper/polymer flexible circuit or tape with metal layers on both sides).

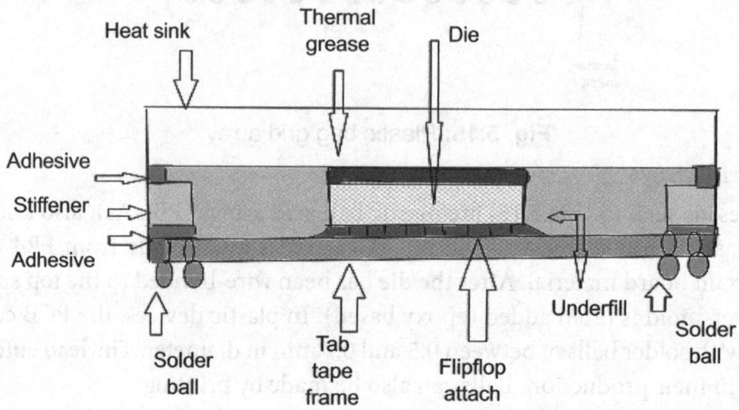

Fig. 5.17: Tape ball grid array (TBGA)

The tape is made from two metal layers: a continuous ground plane on top, and copper lines connecting the die to the solder balls below. There are a few different ways to connect the die to the copper lines, including wire bonding, solder reflow, and thermosonic inner lead bonding. After it's attached, the die is enclosed for safety. Micro-welding, which is similar to wire bonding, connects tiny solder balls to the other end of the copper line. Size-wise, the solder spheres are around 0.8mm in diameter

- **Area array formats**

Full arrays, perimeter arrays, and partial or modified arrays are the three most typical area array configurations for BGA technology [6]. Creating complete-area arrays is a breeze when PBGAs, CBGAs, and CCGAs are used. A PBGA, CBGA, CCGA, or TPGA can be used in a partial or customised array. TPGAs require this array architecture because solder balls cannot reach the die in the package's centre. Because of worries about the solder balls buried beneath the silicon's surface, we're seeing a rise in partial area array layouts. These strains come from the BGA's building materials, which exhibit widely varying CTEs. When a silicon die is being thermally loaded, the high strains caused by the loading can cause the die to detach from the carrier or break if a ball is put close to the edge of the die.

5.3.9 COB (Chip-on-Board)

During the chip-on-board process, a semiconductor is epoxy-attached to a PCB, wire-bonded, and then encased in polymeric materials. The maximum package density and fastest signal speed are achieved through direct wire attachment of the semiconductor to the circuit board.

5.4 PACKAGING OF PASSIVE COMPONENTS AS SMD

It is preferable that all components, not only semiconductors, be surface mounted to make the most of the benefits of surface mount technology. Resistors, capacitors, inductors, and so on are all examples of such parts. Cubic capacitors and resistors, sometimes called "chips," are readily available. Resistors, capacitors, and diodes are the most common electronic components found on chips. However, any device with two ends may be purchased as a chip. Rectangular packages are the most popular, with solderable terminations on one end, both ends, the top and bottom, and often the sides as well. Figure depicts sample metallizations found on a chip component.

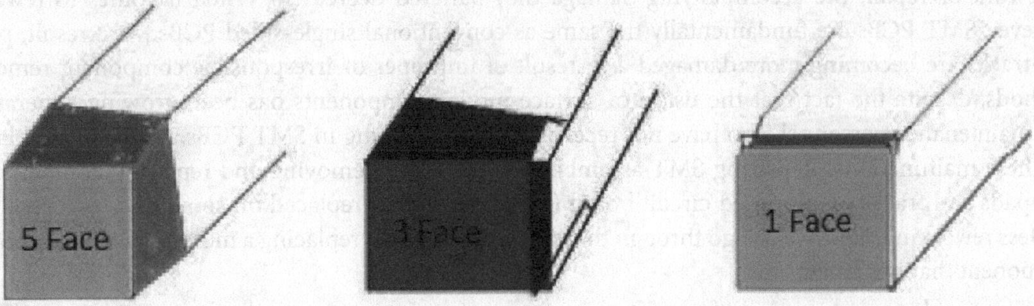

Fig. 5.18: Difference between one face ,three face , five face materialization on chip components

(Fig. 5.19) depicts typical cases for a variety of passive electronic components such capacitors, resistors, and inductors. Surface-mount (SMD) resistors don't have wires connecting them to other components. They're soldered right onto the printed circuit board. The improved heat dissipation properties and consistent resistance values of SMD resistors are the result of their exact production in automated facilities. Power ratings for SMD components are based on their length and breadth.

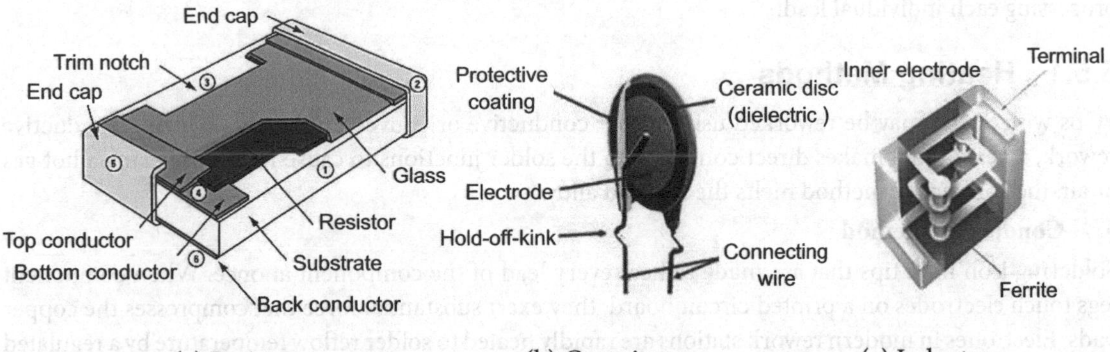

(a) Resistors (b) Capacitor (c) Inductor

Fig. 5.19: Surface mount Passive components

5.5 REPAIRING SURFACE MOUNTED PCBS

It will be unusual to find a company producing products in the next few years that do not have surface mount electronics. As such, the success as an electronics repair firm relies on your possession of both the appropriate equipment for the work and the expertise to properly use those tools. Despite the longevity of surface mount technology, service and repair organisations have been sluggish to adopt the necessary equipment for removing and replacing surface mount devices. Maybe the price was too much for the ordinary individual. When the proper tools weren't available, the technicians would get creative with what they had available, often to the detriment of the whole board spending money on repair and rework stations is warranted because, to do a job befitting your expertise and training, you need access to the same tools and facilities that are used to maintain the expensive equipment. Removing electronic components has become less of a hassle as double-sided printed circuit boards with plated-through holes have replaced single-sided boards in electronic component manufacturing. When PCBs were subjected to rework or repair, the accompanying damage they suffered decreased. When it comes to rework, however, SMT PCBs are fundamentally the same as conventional single-sided PCBs. As a result, pads and tracks are becoming more damaged as a result of improper or irresponsible component removal methods, despite the fact that the usage of surface mount components has been growing. Operators and maintenance personnel who have not received enough training in SMT PCBs are mostly to blame for these malfunctions. Repairing SMT assemblies often entails removing and replacing components. The pads and tracks on a printed circuit board (PCB) need to be replaced on sometimes as a result of careless reworking. Here we will go through the various options for replacing a multi-lead surface mount component that has failed.

The simplest way to get rid of a malfunctioning component is to sever all of its connections. When everything else fails, this is the next best option. The removal method is carefully slicing through each leg in turn. The last IC leg is pulled off with tweezers after each junction is melted with a fine tip, temperature controlled soldering iron.After the solder has had time to cool, it may be removed using a desoldering braid. This approach has the benefit of being low-cost and easy to implement in the field without the need for any specialized equipment. The PCB substrate and copper pads may be harmed in addition to the component. Moreover, when working with fine pitch multi-lead devices, it might be very difficult, if not impossible, to solder the new component in place with a soldering iron since this involves processing each individual lead.

5.5.1 Heating Methods

PCBs with SMDs may be reworked using either conductive or convection heating. During conductive rework, a heated tool makes direct contact with the solder junctions to cause reflow. By using a hot gas or air, the conductive method melts the tin-lead alloy.

- **Conductive method**

Soldering iron have tips that are made to heat every lead of the component at once. When component legs touch electrodes on a printed circuit board, they exert substantial force that compresses the copper pads. Electrodes in modern rework stations are rapidly heated to solder reflow temperature by a regulated current pulse (approximate three seconds).The component may be picked up using the device's built-in vacuum pick-up as the solder on the joints melts. This method permits fast cooling of all leads after soldering, which keeps them in place while the solder hardens. There are several benefits to using this approach. It can be repeated quickly and reliably, and the component body is not heated in the process. Due to the electrodes' ability to keep the legs flush with the pads during solder reflow, this technique is ideal for replacement; alignment and orientation can be checked under a microscope to guarantee success. The high cost and the need for specialised machinery to produce gull-wing (QFPs) and TAB pieces are a few of the drawbacks.

- **Vacuum-Pickup Dual-Heater**

This unique tool is used for removing bigger components. Once the reflow is established, the bigger tips may be removed with one hand thanks to the dual heating system and vacuum pick. All common flat packs and a few BGAs may be easily removed using this tool. With its two heaters and squeezing action, the thermal tweezer can remove components as tiny as chips and as big as PLCCs and leadless packages. Because the tips are brought into close proximity to the solder joints by the tweezer movement, a lot of heat is transferred to the joints at a low temperature.

- **Convective Method (Hot Gas Soldering)**

Hot gases or hot air are often used as the heat transfer medium in production and rework stations. Chips, transistors, SOICs, and flat packs, among other miniature components, are easily extracted using a nozzle with a single, focused point of entry. The leads are warmed with the gas stream until complete reflow occurs, and then the component is removed using tweezers. It takes longer to remove than using a conductive tool, but you just need a single tool and a single nozzle shape for all your needs. To remove any SMDs that may be on just two or three sides of a longer part, a nozzle tailored to that portion is attached to the hand piece and swung around the item. Component lift-off upon reflow is facilitated by the presence of vacuum. It's also possible to reflow the solder junctions with the help of an infrared rework station. Both vacuum pick-up mechanisms and magnifying systems with video display units (VDUs) to permit monitoring of the work in progress are typical examples of supplementary features.

5.5.2 Removal and Replacement of Surface Mount Devices

In order to dismantle a machine with a hot gas machine, you must follow these procedures:

- To all joints, apply a small quantity of liquid flux. Use the proper head for the job.
- Turn on the gas flow while the PCB is in position to reflow the solder at each junction. (confirm with a microscope or visual display unit).
- If the part is held together with glue, you can remove it by turning the head.
- Put the PCB in a cool place and take out the component using the vacuum pick-up.
- Make use of fine desoldering braid to get rid of any stray globs of solder.
- Finally check for any signs of damage to the pads.

The steps here are taken in order to replace the component:

Make sure the replacement part's legs aren't bowed or warped by giving them a thorough inspection. The device's legs should ideally be 12 degrees steeper than the body. When the component is inserted into the PCB, the legs will naturally flatten onto the pads. The pads are given a minimal coating of flux. After making sure the part is value and competitive advantage with the PCB, it is inserted into the hot gas machine's head. The Solder quick tape's aligning fingers will make it easier to place the part in the correct location.

The part must be elevated off the board surface until the pads are just above the component's legs before the gas flow is started. The next step is to turn on the gas glow. In order to melt the solder on the pads and warm up the legs, we need gas. A component should be gently lowered into the board while the solder melts, with the legs of the component centred over the pads and resting between the fingers of solder masking tape. To guarantee proper solder flow around each leg, keep the gas on for a few extra seconds.

In order to protect the connections when you remove the board from the machine, switch off the gas when the solder is flowing easily and let the board cool for at least a minute. Carefully peel off the Solder Quick tape and wipe away any lingering flux from the joints once the PCB has been removed. Use a magnifying glass of at least X10 to check each joint for proper reflow. Clean the PCB using isopropyl alcohol aerosol to get the solvent beneath the component and flush off the flux. After that, you may brush the area to get rid of any remaining flux.

Raised pads on quad flat pack (QFP) layouts are the most frequent source of trouble for surface mount boards. One likely explanation is that workers have a hard time judging whether the solder joints on the device's package's four sides are fully melted. The following procedure is recommended for fixing such damage:

- Clean the area surrounding the damaged pad/track on the board after uninstalling it.
- The replacement track or pad you choose has to work with the original (A variety of companies offer these.).
- Soldering the replacement pad or track to the remaining section of the broken track will connect it.

(Fig. 5.20) depicts the damaged section of track and the new pad that will be used to repair it. The replacement track is laid over the damaged track, and they are then soldered together.(Fig. 5.21).

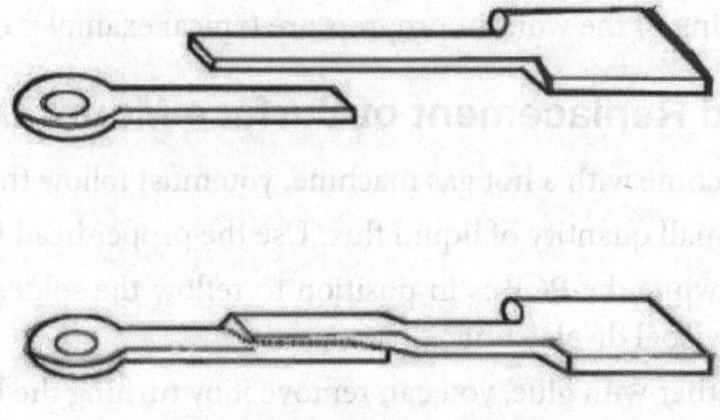

Fig. 5.20: Working repair

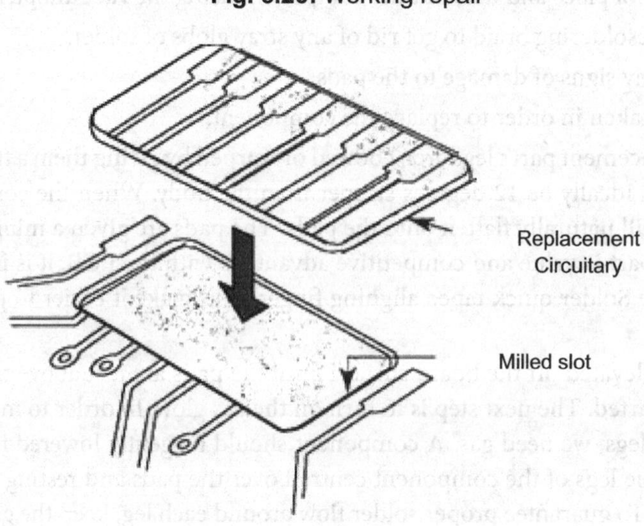

Fig. 5.21: Serviceable repair

Apply a high-temperature glue to the PCB substrate before adding the new pad or track. Closely secure until the glue has set.Insert the new part by soldering it into position,get rid of the flux mess,any solder resist that has been cleaned off should be reapplied.

One may use the metaphor of "steaming a stamp off of an envelope" to describe the process of removing a surface mount component. In practise, this is achieved by heating the solder around the joints of a component and then taking it off the PCB at the same time. After the substrate has been cleaned, a new part may be soldered onto the board. Choosing the right technique for heating the solder and component leads is important. There are a variety of implements that may be used for this, including those with conductive or convective properties, as well as those with single or many points, attainable or not. In order to replace a faulty part in a through-hole assembly, the assembly must be desoldered. Before a new component can be installed on the board, the old one must be carefully removed, the hole must be cleaned, and the solder must be removed. A surface mount board does not need soldering. In contrast, molten solder is taken from the lead and plated through-hole (PTH) on the through-hole board using a vacuum. All leads on a through hole may be re-flowed at once using a hot air tool or solder pot, enabling the component to be removed. All of a component's leads need to be heated simultaneously in order to remove it from a surface mount board before the part hardens once again. The footprint on the board can be damaged if not all leads are heated at once and the device is removed before all the solder has melted. This could cause issues with the coplanarity of the new component on the PCB. Manufacturers' increased adoption of fine pitch technologies in surface mount design has made rework a more challenging operation. With a smaller board pitch, component misalignment and PCB heat deterioration are more obvious. When redoing fine pitch boards, a vision system is often required. Having a vision system that allows for concurrent viewing of the PCB and component is becoming increasingly important as lead count decreases. Fine pitch components need the use of optical equipment for installation to guarantee accurate alignment. The ideal rework station would include the following features, taking into account:

- A sighting system for positioning and soldering the component
- To have a placement tool capable of modifications even more minute than the smallest pitch presently applied to the board.
- It needs to be easy to use, provide heating control, and heat the board and component in a manner that is consistent with the original manufacturing method. When replacing or removing components, it must be able to deliver homogeneous heat without laminating the board or damaging the component.

5.6 REWORK STATIONS

Precision and performance for modern printed circuit boards, which may include BGAs, DCAs, CSPs, and fine pitch SMDs, cannot be guaranteed with hand-held tools only. Area array components make rework more difficult. The bumps' location on the chip's bottom makes it challenging to align and inspect the inter-connections with the pads, which can mask voids, bridges, and other problems until functional testing reveals them. Soldering iron tip temperature, duration at each pad, applied pressure, impacted area, contact area, and placement are only some of the factors that must be controlled while desoldering manually with a wick. However, vacuum desoldering tools have a number of factors that must be managed precisely. These include the vacuum flow, the distance from the pad, the hot air flow (if necessary), the hot air temperature, and the source pressure. How well the operator controls these parameters determines how much damage will be done to the pads, traces, and solder marks. Assuring high quality and low overhead in rework processes is possible with the help of a technician-free automated work station.

Rework tools come in a huge range right now. Typical rework stations may be found in electronic repair businesses, and the SD-3000 from M/s Howard Electronic Instruments, USA, is a good illustration of a popular model. The SMD is heated using a single nozzle on microprocessor-controlled equipment that blows hot air in a tracing motion along the solder joints. This apparatus may be used to reflow and/or remove QFP, SOP, PLCC, PGA, BGA, etc. components of any size or form (solder). With no need to swap out the nozzle head, it can process any SMD. The PCB and neighbouring components are protected from overheating thanks to an integrated timer.

The equipment's many buttons and dials are seen in(Fig.5.22) :

- X-Axis: This variable allows you to modify the nozzle's width in relation to the size of the component being re-flowed. Moreover, it serves as the internal adjustment while removing BGAs and PGAs.
- Y-Axis: A component's nozzle length is modified using this knob before it is reflowed. It is also the external adjustment while removing BGAs and PGAs.
- Z-Axis: The Z-Axis slider sets the nozzle's distance from the solder joints that need reflowing.
- Nozzle: Knobs on the X, Y, and Z axes allow you to position the nozzle so that it rotates around the solder points of the component being removed. The heater's temperature is indicated by the colour of the heater coils, which may be inspected through holes in the nozzle.
- Timer: When the start button is pressed, the timing is adjusted such that it takes only long enough to achieve the solder melt temperature. The device enters a cool down cycle and turns off once it reaches the predetermined temperature atthe end of the time period.
- Temperature: This dial adjusts the heater's temperature in any mode of operation to suit the operator's tastes.
- Mode: Eliminating BGA/PGA, QFP/PLCC standard packages is made possible by the mode switch. High air velocity (12 l/min) and temperature are other features of this mode. The airflow rate is reduced to 6 l/min in the LOW setting. For densely populated boards, this mode is utilised to prevent the inadvertent removal of tiny chips during the re-flow process. Once the old solder has been cleaned from the pads and new solder paste has been put to the new component to be reflowed, you can go to SLOW mode and interchange the QFP/PLCC packages.
- Start: As long as you press the start button, the nozzle will begin to rotate, allowing you to change its length and breadth. Once the settings have been adjusted to your satisfaction, press the start button a second time to activate the fan, heater, and timer.
- Stop: If you need to remove a component from a printed circuit board while utilising a vacuum, you can do it by pressing the stop button, which will immediately turn off the heating element and lift the nozzle by around half an inch.

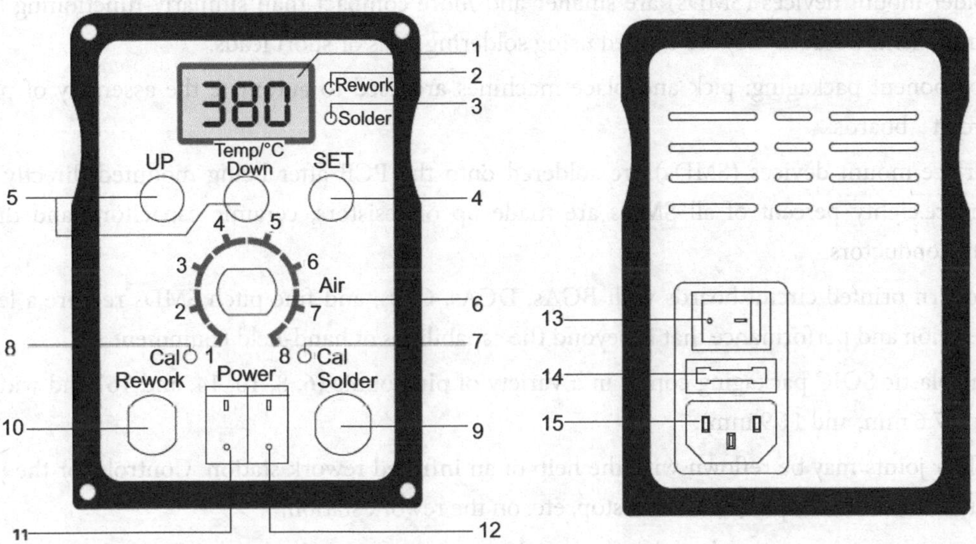

Fig. 5.22: Controls on typical rework station

A vacuum picks up the IC to be removed from the board, and a mechanism holds the PCB steady while you slide it under the hot air nozzle for component alignment. The following can help you achieve shorter reflow durations and cooler temperatures while making repairs to QFP or PLCC components:

- Keep the nozzle at a maximum height of 2 mm above the tabletop at all times. The heater head/ nozzle assembly and the circuit board may need to be mounted in a fixture for this purpose.
- Increase the air flow as much as you can without causing the solder connections on the edges to melt.
- Toss in some flux By following these steps, the technician will be equipped with the necessary connective tools and knowledge of the effects on assemblies to build his or her own rework process, should that be necessary.

Know more about Rework station

UNIT SUMMARY

- Surface mount technology is steadily replacing the traditional through-hole mounting method for printed circuit components.
- Advantages of surface mounting include streamlined manufacturing, smaller circuit boards, and higher dependability.
- Due to ongoing demand and the market trend toward downsizing, surface mount technology (SMT) has been introduced to electronic assembly in order to boost density while decreasing board area.

- Solder-mount devices (SMDs) are smaller and more compact than similarly-functioning leaded components because they are created using soldering pads or short leads.
- Component packaging: pick and place machines are used to automate the assembly of printed circuit boards.
- Surface-mount devices (SMDs) are soldered onto the PCB after being mounted directly to its surface.eighty percent of all SMDs are made up of resistors, ceramic capacitors, and discrete semiconductors.
- Modern printed circuit boards with BGAs, DCAs, CSPs, and fine pitch SMDs require a level of precision and performance that is beyond the capabilities of hand-held equipment.
- The plastic SOIC packaging comes in a variety of pin counts (6, 8, 10, 14, and 16) and widths (4 mm, 7.6 mm, and 11.9 mm).
- Solder joints may be reflown with the help of an infrared rework station. Controls for the y-axis, z-axis, nozzle, temperature, start, stop, etc. on the rework station.
- Damage to surface mount boards often manifests as elevated pads on quad flat packs (QFP)
- **The through-hole board is distinguished from the surface-mount board by the use of a vacuum to draw out the molten solder from the lead and plate the through-holes.**
- **When dismantling a large part, the Dual Heater with Vacuum Pick-up is the tool of choice. The dual heating system and the integrated vacuum pick allow the larger tips to be removed with one hand when the reflow has been established.**

EXERCISES

Multiple Choice Questions

1. What is the technology used to mount the components directly.

   (a) Surface Mount Technology (b) Sensor Moulding Technology
   (c) Silicon Made Technology (d) Surface Movement Technology

2. Solder joint in Surface Mount Technology forms _____?

   (a) Electrical Connection (b) Mechanical Connection
   (c) Chemical Connection (d) Both A and B

3. What is not used to connect in Surface Mount Technology?

   (a) Holes (b) Terminals
   (c) Both (a) and (b) (d) Surface

4. Surface Mount Capacitors utilized for what?

   (a) Decoupling Applications (b) Frequency Control
   (c) Both (a) and (b) (d) None

5. What is the dielectric for the capacitors manufactured from Surface Mount Technology?
   (a) Ceramic (b) Tantalum
   (c) Either (a) or (b) (d) Aluminum
6. What type of Surface Mount Capacitors offer higher Volumetric efficiencies?
   (a) Surface Mount Tantalum Capacitors
   (b) Surface Mount Aluminum Capacitors
   (c) Surface Mount Ceramic Capacitors
   (d) Surface Mount Germanium Capacitors
7. How are Surface Mount Devices de-soldered?
   (a) Soldering Iron (b) Solder Wire
   (c) Hot Air Blower (d) Solder Paste
8. What are the various types of Surface Mounting?
   (a) Type-1 (b) Type-2
   (c) Type-3 (d) All Mentioned Above
9. Which type of Surface Mount Technology consists of Discrete components only?
   (a) Type-1 (b) Type-2
   (c) Type-3 (d) All Mentioned Above
10. In type 3, the Surface Mount Components are glued at_?
   (a) Top Side (b) Bottom Side
   (c) Both Sides (d) None
11. Type 2, assembly is the combination of_______?
   (a) Type-1 (b) Type-3
   (c) Both A and B (d) Discrete Components
12. Assembling in type 1 SMT is_?
   (a) Single Sided (b) Double Sided
   (c) Either A or B (d) Multiple Sides
13. In type 2 SMT______doesn't present on bottom?
   (a) Passive Surface Mount Devices (b) Active Surface Mount Devices
   (c) Surface Mount Capacitors (d) Surface Mount Resistors
14. What increases the complexity of SMT assembling?
   (a) Large Fine Pitch (b) Ultra Fine Pitch
   (c) Quad Fl (d) All above
15. What are the various Surface Mount Technology Devices?
   (a) Lead-less Ceramic Chip Carriers (b) Dual In-Line Package
   (c) Plastic Lead Chip Carrier (d) All Mentioned

ANSWERS															
1.	(a)	2.	(d)	3.	(d)	4.	(c)	5.	(c)	6.	(a)	7.	(c)	8.	(d)
9.	(c)	10.	(b)	11.	(c)	12.	(c)	13.	(b)	14.	(c)	15.	(c)	10.	(c)

Short Answer and Long Answer Type Question

Category 1

1. Define Surface mount technology.
2. What are the Benefits of SMT?
3. What is the Process Used in SMT Assembly?'
4. What are the Different Testing Services Done During SMT Assembly
5. How is SMT assembly different from THT assembly?
6. Why is SMT assembly widely applied in electronics manufacturing?
7. What is chip mounting and its role in SMT assembly?
8. Why type of soldering is used in SMT assembly procedure?
9. What are various Surface Mount Resistors?
10. What is Tape ball grid array

Category 2

1. Analyze the Flat packs and Quad Packs
2. Explain the Rework station and summarize its control panel
3. What is PLCC explain the need of PLCC.
4. Describe the packaging of passive component as surface mount device
5. Differentiate between Land grid array and Ball grid array

REFERENCES AND SUGGESTED READINGS

1. Tummala R R, Rymaszewski E J and Klopfenstein A G:'Microelectronic Packaging Handbook'. Chapman and Hall, 1997.
2. Heitmann R:'The ultimate connections: BGA and flip chip attachment'. Electronic Packaging and Production, May 1996, 68-76.
3. Tarter T S, Goetz M P and Pagageorge BGA performance characteristics: a users design guide'. Proc of Tech Programme, SMI, 1995, 245-254.
4. Cole M S et al,'Shock and Vibration limits for CBGA and CCGA Proceedings of the Technical Program, SMI, CA Sept 1996, 89-94
5. Harvey P M:'Electrical Characterization of Tape Ball Grid Array Packaging'. Area Array Packaging Technology Workshop, Nov 1995, Berlin.
6. Khandpur, R.S. (1996), Handbook of Analytical Instruments, Tata McGraw- Hill Publishing Co. Ltd., New Delhi. ISBN 0- 07-460186-5
7. Loveday, G (1989), Electronic Testing and Fault Diagnosis, Longman (Scientific and Technical), England. ISBN -582- 03865-0

8. Malvino, A.P. (1989), Electronic Principles, 4th ed. McGraw-Hill, U.S.A. ISBN 0-07099479
9. https://archive.nptel.ac.in/courses/112/105/112105232/
10. http://www.nitttrc.edu.in/nptel/courses/video/101104075/lec18.pdf
11. https://www.youtube.com/watch?v=mKjSXPXVyRk

KNOW MORE

SMT stands for Surface Mount Technology, the entire technology of mounting and soldering electronic components, similar as resistors, capacitors, transistors, integrated circuits, onto a printed circuit board or PCB. The factors used are also applied to as surface mount devices(SMD, surface- mount devices). It should be noted that SMT doesn't need to reserve corresponding through holes for component pins, and SMD is much smaller than through- hole insertion technology.

QR SCAN CODE FOR SUPPORTIVE KNOWLEDGE EMBEDDED IN THE CHAPTER

❑❑

APPENDICES

APPENDIX-A

Resistors colour code and characteristic

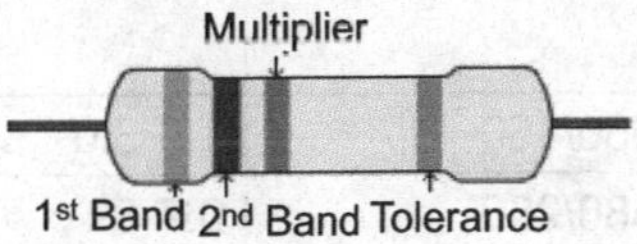

COLOUR	DIGIT
Black	0
Brown	1
Red	2
Orange	3
Yellow	4
Green	5
Blue	6
Violet	7
Gray	8
White	9

COLOUR	TOLERANCE
Gold	± 5%
Silver	± 10%
No colour	± 20%

Resistors Types

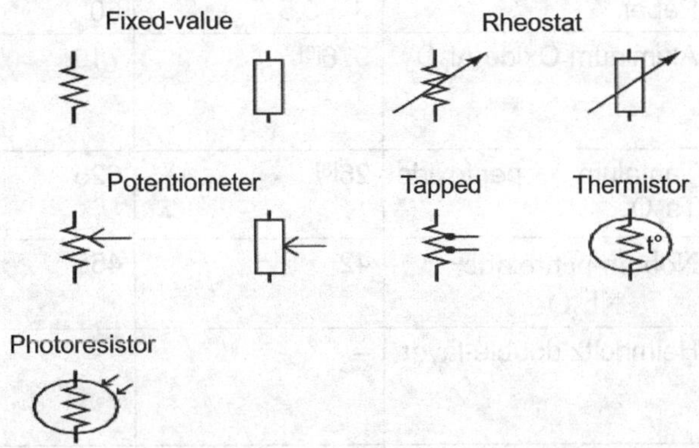

APPENDIX-B

Types of capacitor with their characteristics

Capacitor style	Dielectric	Permittivity at 1 kitz	Maximum/ realized. dielectric strength V/µm	Minimum thickness of the dielectric µm
Ceramic capacitors Class 1	paraelectric	12-40	<100(7)	1
Ceramic capacitors, Class 2	termoelectric	200-14.000	<25(7)	0.5
Film capacitors	Polypropylene (PP)	2.2	650/450	1.9-3.0
Film capacitors	Polyethylen terephthalate, Polyester (PET)	3.3	580/280	0.7-09
Film capacitors	Polyphenylene sulfide (PPS)	3.0	470/220	1.2
Film capacitors	Polyphenylene nephthalate (PEN)	3.0	500/300	0.9-1.4
Film capacitors	Polytetrafluoroethylene (PTFE)	2.0	450(?)/250	5.5
Paper capacitors	Paper	3.5-5.5	60	5-10
Aluminum electrolytic capacitors	Aluminum Oxide Al_2O_3	9, 6[8]	710	<0.01 (6.3 V) <0.8 (450 V)
Tantalum electric capacitors	Tantalum pentoxide Ta_2O_5	26[8]	625	<0.01 (6.3 V) <0.8 (450 V)
Nobum electrolytic capacitors	Nobum pentoxide Nb_2O_5	42	455	<0.01 (6.3 V) <0.10 (40 V)
Supercapactiei Double-layer capactors	Helmholtz double-layer	—	—	<0.001 (2.7 V)
Vacuum capacitors	Vaccum	1	40	—
Air gap capactors	Air	1	3.3	—
Glass capactors	Glass	5-10	450	—
Mica capacitors	Mica	5-8	118	4-50

Capacitor type	Dielectric	Features/applications	Disadintages
Ceramic Class 1 capacitors	**paraelectric** ceramic mixture of **Titanium** dioxide modified by additives	Predictable linear and low capacitance change with operating temperature. Excellent high frequency characteristics with low losses. For temperature compensation in resonant circuit application. Available in Voltages up to 15,000 V	Low **permittivity** ceramic, capacitors with low **volumetric efficiency**, larger dimensions than Class 2 capacitors
Ceramic Class 2 capacitors	**ferroelectric** ceramic mixture of **barium titanate** and sutable additives	High permittivity high volumetric efficiency, smaller dimensions than Class 1 capacitors. For buffer, by-pass and coupling applications. Available in voltages up to 50.000 V	Lower stability and higher losses than Class 1. Capacitance changes with change in applied voltage, with frequency and with aging effects. Sighty **microphonic**

Capacitor Types

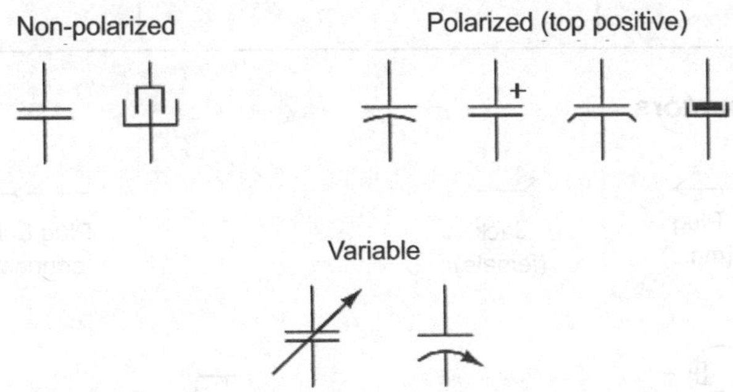

APPENDIX-C

Electrical Symbol

Convention used in this book

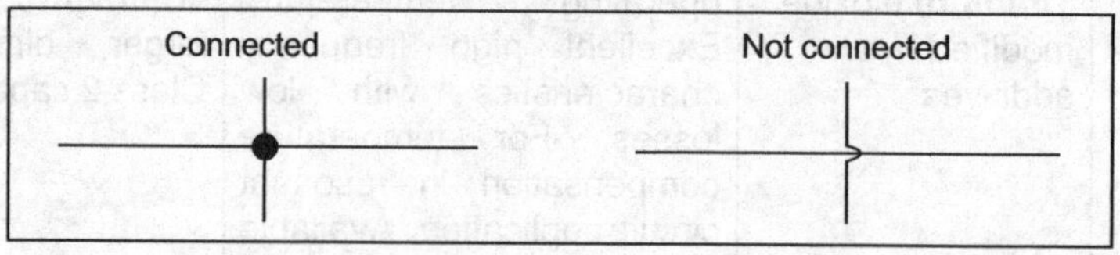

Older convention

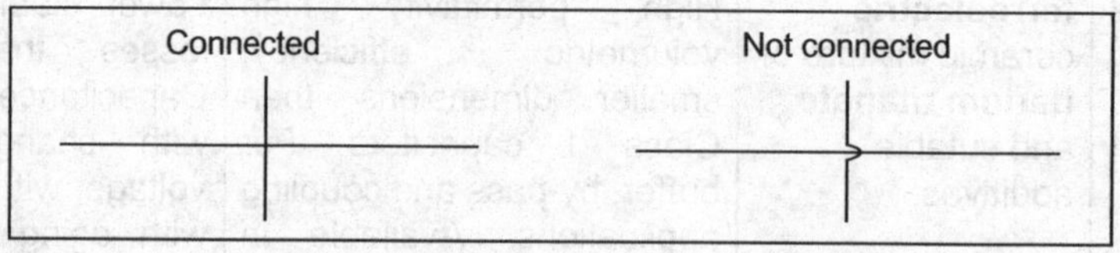

Newer convention

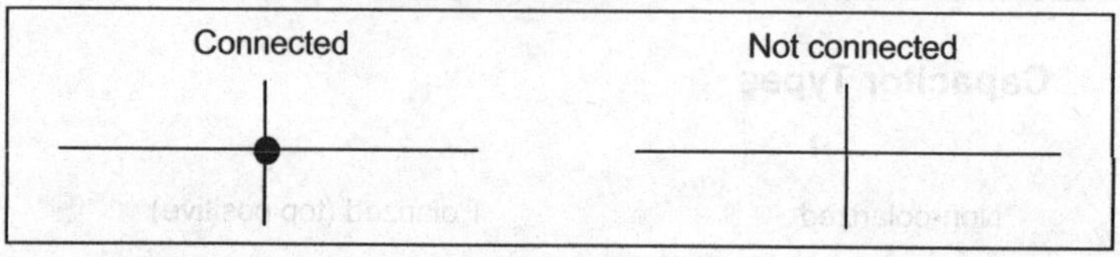

Connectors

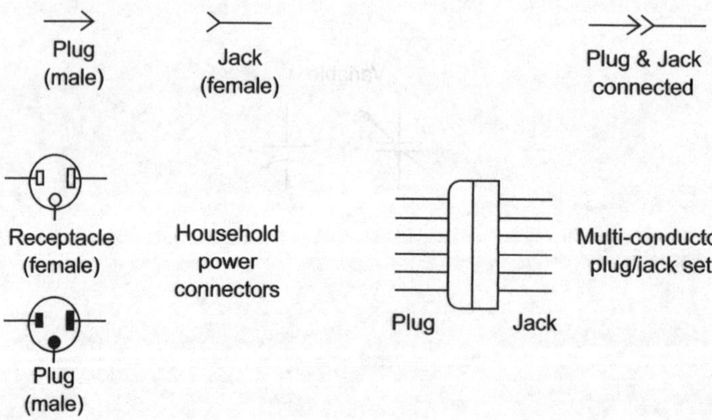

Switches, Electrically Actuated (Relays)

Relay components, "ladder logic" notation style

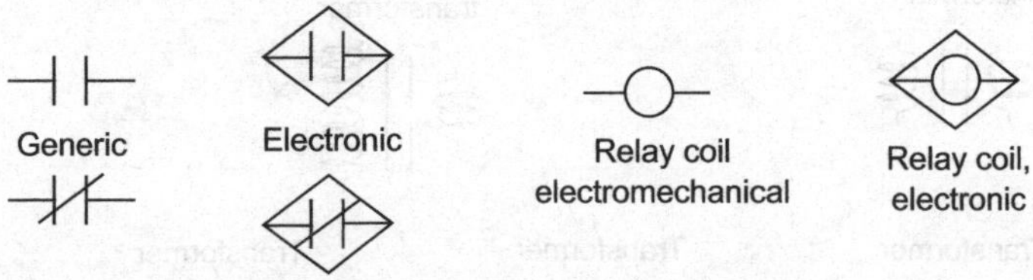

Rolay, electronic schematic notation style

Switches, Hand Actuated

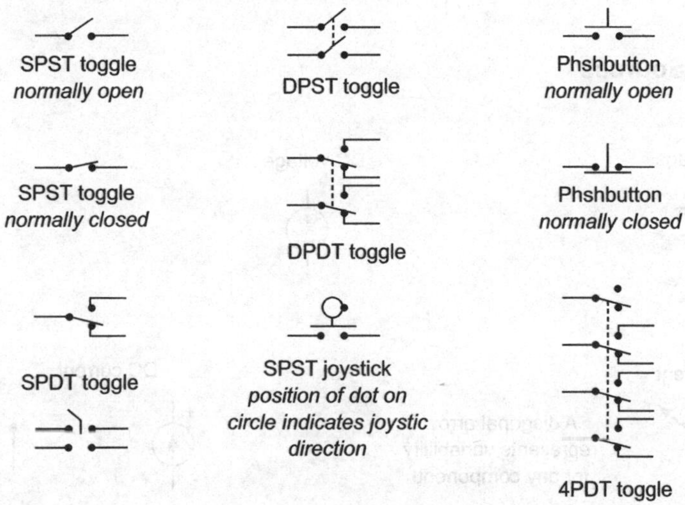

Mutual Inductors

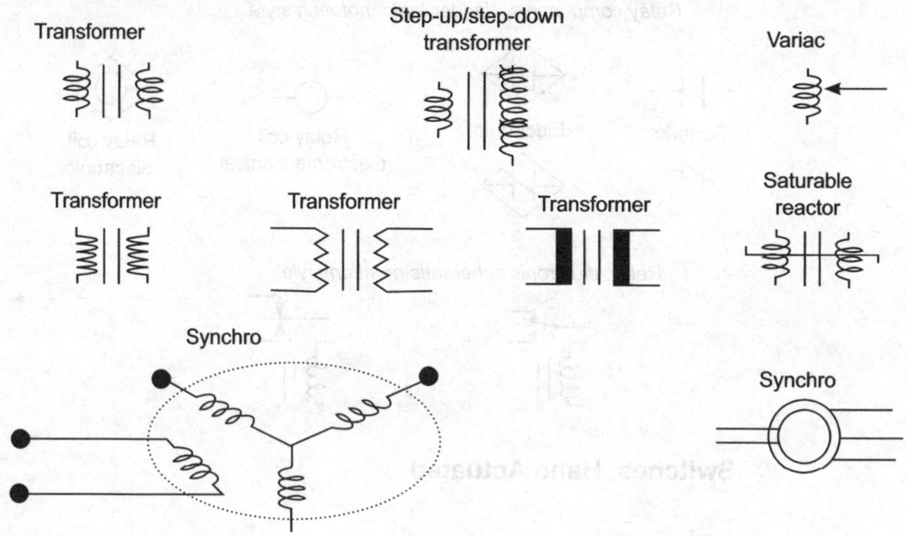

Power Sources

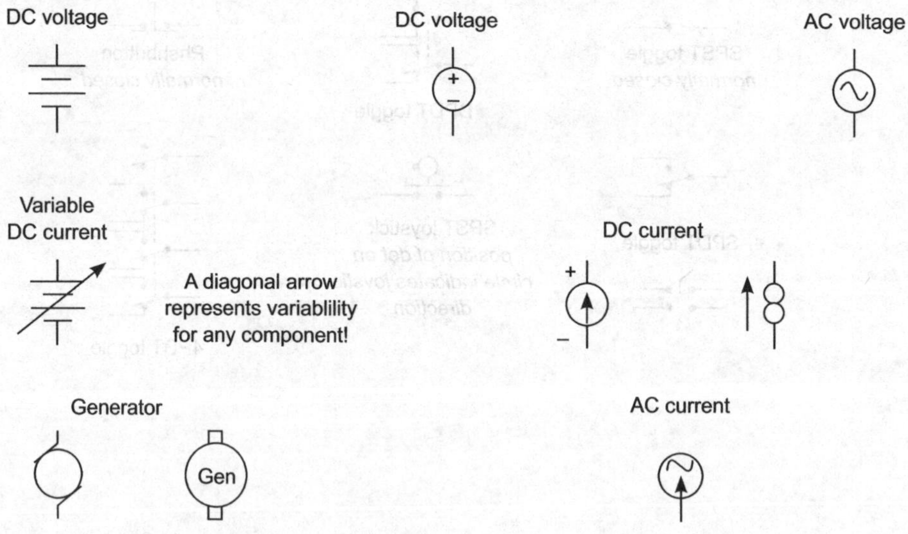

CO AND PO ATTAINMENT TABLE

Course outcomes (COs) for this course can be mapped with the programme outcomes (POs) after the completion of the course and a correlation can be made for the attainment of POs to analyze the gap. After proper analysis of the gap in the attainment of POs necessary measures can be taken to overcome the gaps.

Table for CO and PO attainment

Course Outcomes	Expected Mapping with Programme Outcomes (1- Weak Correlation; 2- Medium Correlation; 3- Strong Correlation)						
	PO-1	PO-2	PO-3	PO-4	PO-5	PO-6	PO-7
CO-1	3	3	1	3	2	2	3
CO-2	2	3	2	3	2	1	2
CO-3	1	2	3	3	2	1	2
CO-4	2	3	2	3	2	2	3
CO-5	2	3	2	3	3	2	2

The data filled in the above table can be used for gap analysis